ADVANCES IN MEDICINE AND BIOLOGY

VOLUME 148

ADVANCES IN MEDICINE AND BIOLOGY

Additional books and e-books in this series can be found
on Nova's website under the Series tab.

ADVANCES IN MEDICINE AND BIOLOGY

VOLUME 148

LEON V. BERHARDT
EDITOR

Medicine & Health

New York

We have partnered with Copyright Clearance Center to make it easy for you to obtain permissions to reuse content from this publication. Simply navigate to this publication's page on Nova's website and locate the "Get Permission" button below the title description. This button is linked directly to the title's permission page on copyright.com. Alternatively, you can visit copyright.com and search by title, ISBN, or ISSN.

For further questions about using the service on copyright.com, please contact:
Copyright Clearance Center
Phone: +1-(978) 750-8400   Fax: +1-(978) 750-4470   E-mail: info@copyright.com.

NOTICE TO THE READER

The Publisher has taken reasonable care in the preparation of this book, but makes no expressed or implied warranty of any kind and assumes no responsibility for any errors or omissions. No liability is assumed for incidental or consequential damages in connection with or arising out of information contained in this book. The Publisher shall not be liable for any special, consequential, or exemplary damages resulting, in whole or in part, from the readers' use of, or reliance upon, this material. Any parts of this book based on government reports are so indicated and copyright is claimed for those parts to the extent applicable to compilations of such works.

Independent verification should be sought for any data, advice or recommendations contained in this book. In addition, no responsibility is assumed by the Publisher for any injury and/or damage to persons or property arising from any methods, products, instructions, ideas or otherwise contained in this publication.

This publication is designed to provide accurate and authoritative information with regard to the subject matter covered herein. It is sold with the clear understanding that the Publisher is not engaged in rendering legal or any other professional services. If legal or any other expert assistance is required, the services of a competent person should be sought. FROM A DECLARATION OF PARTICIPANTS JOINTLY ADOPTED BY A COMMITTEE OF THE AMERICAN BAR ASSOCIATION AND A COMMITTEE OF PUBLISHERS.

Additional color graphics may be available in the e-book version of this book.

Library of Congress Cataloging-in-Publication Data

ISBN: 978-1-53616-181-6
ISSN: 2157-5398

Published by Nova Science Publishers, Inc. † New York

CONTENTS

PREFACE

Advances in Medicine and Biology. Volume 148 begins by summarizing recent findings in exosome research, highlighting the importance of exosomes as novel biomarkers and regulatory elements in the process of angiogenesis. Specifically, the authors address the potential of exosomes as future diagnostic and therapeutic tools.

Additionally, the authors discuss the latest findings on myometrial cells embryogenesis, their later phenotypic changes, and structure and its relation to contractility, particularly focusing on the least understood aspects of the myometrium function.

Next, overview of new developments in breast cancer targeted therapies is provided, and future prospects for directed therapeutic strategies are discussed. Recent advances in targeted therapy have decreased the morbidity and enhanced the quality of life of cancer patients.

Several strategies to combat H1N1 through in silico approach along with the discovered drug candidates are reviewed, such as neuraminidase inhibitors, M2 proton channel blocker, and RNA polymerase-complex inhibitor to interfere with H1N1 virus replication. H1N1, also recognized as Swine Flu, is a subtype of influenza A virus which belongs to the orthomyxovirus category.

This compilation also explores acute kidney injury, a life-threatening situation which has a mortality rate of up to 50%. Histone deacetylase inhibitors were first used for treating different types of cancers, but recently their immunomodulatory and anti-apoptotic effects in renal cells in vitro have been noticed. Thus, the effects of histone deacetylase inhibitors in preventing acute kidney injury are reviewed.

The closing paper reviews the literature regarding how to obtain an optimal anatomical reduction in femoral neck fractures in patients under 65 years old, and functional results are discussed in relation to specific fixation implants, fracture patterns and patient-related factors.

Chapter 1 - Current research has been focused on the role extracellular vesicles (EVs) play in different physiological and pathophysiological conditions. Exosomes, the subset of small EVs (30-150 nm in diameter) originating from the endocytic compartment of the parent cells, are of special interest. Exosomes are produced by many different cell types and are considered to be essential contributors to intercellular communication. Exosome importance in angiogenesis is being intensely investigated in health and disease. Various aspects of exosome contributions to angiogenesis have recently been uncovered, including the relevant molecular and genetic exosome cargos, their interactions with endothelial cells and their impact on endothelial cell function. Still, many questions remain unanswered. The published research mainly considers the roles of exosomes in tumor angiogenesis or angiogenesis in the cardiovascular system, especially the heart. Attention has been given to mesenchymal stromal cell-derived exosomes and their beneficial effects in different diseases and to exosomes derived from endothelial cells or endothelial progenitor cells. The objective of this chapter is to summarize recent findings in exosome research, highlighting the importance of exosomes as novel biomarkers and regulatory elements in the process of angiogenesis. Specifically, the authors will address the potential of exosomes as future diagnostic and therapeutic tools.

Chapter 2 - Uterine myocytes provide the vital function - maintenance of a certain tone and quiescence during gestation and strong repeated contractions in labour. To perform such a complicated pattern of activity myometrial smooth muscle cells possess a range of highly specialized structural features that originate in their embryogenesis. The function of uterine myocytes to a great extent depends on their ability to respond to the different regulatory signals (neural, hormonal, mechanical, local) and consequently alter their membrane potential and ion permeability – the basis of excitation-contraction coupling. The problem of preterm labour as a result of incorrect regulation or improper functioning of myometrial cells is still outstanding. 10% of all newborns remain premature and require complicated and expensive care during first months and later often show reduced quality of life. The rate of neonatal death is also higher in the premature group. On the other hand, weak irregular contraction of the myometrium known as uterine hypocontractility is the most common labour complication that could lead to foetal distress and/or postpartum haemorrhage. Thus, investigation of uterine myocytes development, structure and function remains an important area of research. This review covers the latest findings on myometrial cells embryogenesis, their later phenotypic changes, structure and its relation to contractility, which is considered to be the main function of these cells. In particular, the authors focus on one the least understood aspects of the myometrium function, namely the myogenic mechanisms of myocyte contraction. A significant part of the chapter therefore outlines recent data about the role of mechanosensitive membrane ion channels in the regulation of contractility. Among these, special attention is given to Transient Receptor Potential channels (TRP), which are known to be significant part in the regulation of smooth muscle contraction. The mechanism of calcium related relaxation will also be outlined, as well as the role of calcium sensitisation in uterus contractility. This review chapter would be interesting for students, scientists and clinicians.

Chapter 3 - Recent advances in targeted therapy have decreased the morbidity and enhanced the quality of life of cancer patients. The use of peptides and monoclonal antibodies as targeted therapies has received considerable attention in recent years, establishing this mode of treatment as an important therapeutic strategy. Breast cancer is the most common type of cancer in women and one of the few malignancies in which tumour heterogeneity have been successfully used for molecular classification and therapeutic intervention. Advances in molecular biology including genomics, epi-genomics and transcriptomics have resulted in the identification of better treatments that have been tailored to target specific pathophysiology. Here the authors provide an overview of new developments in breast cancer targeted therapies and discuss future prospects for directed therapeutic strategies.

Chapter 4 - H1N1, also recognized as Swine Flu, was disseminated first in 1918 and attributed as a pandemic by the World Health Organization in 2009. H1N1 is a subtype of influenza A virus which belongs to the *orthomyxovirus* category. It encodes eleven proteins: envelope proteins (haemagglutinin and neuraminidase), matrix proteins (M1 and M2), Viral RNA Polymerases (PB2, PB1, PB1-F2, PA, and PB), and nonstructural proteins (NS1 and NS2). Neuraminidase and haemagglutinin are the proteins that make a difference on H1N1 strain with other strain of Influenza A. While all of the proteins have a different role, several antiviral compounds to inhibit H1N1 activity have been discovered through *in silico* drug discovery. Computational simulation and bioinformatics have been favoured for years as reliable approach to identify and design a potent drug candidate. Molecular docking and molecular dynamics simulation are useful tools to perform high-throughput screening from various databases to search a novel candidate for H1N1 inhibitor. In this chapter, several strategies to combat H1N1 through *in silico* approach along with the discovered drug candidates are reviewed, such as neuraminidase inhibitors, M2 proton channel blocker, and RNA polymerase-complex inhibitor to interfere with H1N1 virus replication.

Chapter 5 - Acute kidney injury (AKI) is a life-threatening situation which has a mortality rate of up to 50 percent. DNA damage, oxidative stress, apoptosis of tubular epithelial cells, and some other cell-mediated processes are the most common mechanisms involved in AKI. Histone deacetylase (HDAC) inhibitors are a new group of medications which were first used for treating different types of cancers all around the world. However, recent in vitro studies of HDAC inhibitors demonstrated significant immunomodulatory and anti-apoptotic effects of this group in renal cells which could be beneficial in the management of AKI. Since preventing such a dangerous situation seems vital, this article reviews the effects of HDAC inhibitors in preventing AKI. The studies on the effect of inhibition of HDAC1 gene expression in renal interstitial fibroblasts and tubular epithelial cells indicate that HDAC1 enzyme has a huge role in the activation of myofibroblasts and epithelial cells. Moreover, HDAC inhibitors can prevent TGFβ-1 induced fibrogenesis and can exert anti-fibrotic effects in the kidney. Down-regulation of collagen gene expression by acetylation of histones causes an increase in expression of inhibitors of DNA binding/differentiation 2 (Id2), Bone morphologic protein 7 (BMP-7), and E-cadherin; and this process causes a significant improvement in the kidney, bone, and brown adipose tissues. Also, HDAC2 gene suppression decreases the expression of fibronectin and α-SMA. In general, HDAC inhibitors not only can suppress pathologic genes but also have various positive effects on damaged tissues in comparison to healthy tissues. There is various evidence showing that HDAC inhibitors can have preventive and even therapeutic effects in AKI patients. Since preventing AKI seems vital, this review article focuses on the effects of HDAC inhibitors in preventing AKI.

Chapter 6 - Femoral neck fractures in patients under 65 years old are an uncommon injury often caused by a high-energy trauma, with possible fracture comminution and disruption of blood supply to the femoral head. Garden and Pauwels classification systems are currently used to decide the appropriate treatment; on the other hand, the literature shows controversies among surgeons' decision concerning specific treatment variables such as time to surgery, the role of capsulotomy and fixation methods. Hip arthroplasty is not necessarily the first option in young patients because it may not last as long as in low-demand patients. The main goal in such subgroup of patients is to obtain anatomical reduction and a stable fixation to attempt femoral head preservation: it has been demonstrated that specific measurements of the femoral head posterior tilt and alignment are reliable predictors of reoperation. The clinical outcomes can be improved with good pre-operative planning, optimization of surgical procedures and introduction of new improved implants and techniques. Despite this, a femoral fracture in a young patient is a major adverse event: the healed femur would never be the same as the healthy one, and further surgeries might be needed over the years because of the long life-expectancy. Therefore, a fully comprehension of the anatomical and mechanical aspects of the hip appear to be essential. This paper reviews the literature about how to obtain an optimal reduction and how to maintain it through different fixation techniques; furthermore, the functional results will be discussed in relation to the specific fixation implants, to the fracture pattern and to the patient-related factors. Eventually, factors influencing good fracture healing and how to prevent major complications (shortening in comminute fractures, non-union and avascular necrosis of the femoral head) will be discussed.

In: Advances in Medicine and Biology   ISBN: 978-1-53616-181-6
Editor: Leon V. Berhardt   © 2019 Nova Science Publishers, Inc.

Chapter 1

EXOSOMES: NOVEL MEDIATORS OF ANGIOGENESIS IN HEALTH AND DISEASE

*Nils Ludwig[1,2], Miroslaw J. Szczepanski[3,4] and Theresa L. Whiteside[1,2,5],**

[1]Department of Pathology, University of Pittsburgh
School of Medicine, Pittsburgh, PA, US
[2]UPMC Hillman Cancer Center, Pittsburgh, PA, US
[3]Chair and Department of Biochemistry,
Medical University of Warsaw, Warsaw, Poland
[4]Department of Otolaryngology, Centre of Postgraduate
Medical Education, Warsaw, Poland
[5]Departments of Immunology and Otolaryngology,
University of Pittsburgh School of Medicine, Pittsburgh, PA, US

ABSTRACT

Current research has been focused on the role extracellular vesicles
(EVs) play in different physiological and pathophysiological conditions.

* Corresponding Author's E-mail: whitesidetl@upmc.edu.

Exosomes, the subset of small EVs (30-150 nm in diameter) originating from the endocytic compartment of the parent cells, are of special interest. Exosomes are produced by many different cell types and are considered to be essential contributors to intercellular communication. Exosome importance in angiogenesis is being intensely investigated in health and disease. Various aspects of exosome contributions to angiogenesis have recently been uncovered, including the relevant molecular and genetic exosome cargos, their interactions with endothelial cells and their impact on endothelial cell function. Still, many questions remain unanswered. The published research mainly considers the roles of exosomes in tumor angiogenesis or angiogenesis in the cardiovascular system, especially the heart. Attention has been given to mesenchymal stromal cell-derived exosomes and their beneficial effects in different diseases and to exosomes derived from endothelial cells or endothelial progenitor cells. The objective of this chapter is to summarize recent findings in exosome research, highlighting the importance of exosomes as novel biomarkers and regulatory elements in the process of angiogenesis. Specifically, we will address the potential of exosomes as future diagnostic and therapeutic tools.

Keywords: exosomes, angiogenesis, extracellular vesicles, cell communication

INTRODUCTION

The survival of mammalian cells is dependent on oxygen and nutrients. Since the diffusion limit for oxygen is only 100 to 200 μm, the growth of multicellular organisms is dependent on the recruitment of blood vessels (Carmeliet and Jain 2000). Angiogenesis is the process of forming new blood vessels from pre-existing vessels. Angiogenesis plays an essential role in numerous biological processes. Besides its physiological roles in embryogenesis, organ development and wound healing, angiogenesis contributes to the progression of cancer and various ischaemic and inflammatory diseases, such as psoriasis, arthritis and diabetic retinopathy (Carmeliet 2003). Angiogenesis is regulated by a wide spectrum of morphogenic and molecular features, and research in the field has identified numerous pro- and anti-angiogenic factors that contribute to vessel formation and growth (Döme et al. 2007; Carmeliet and Jain 2000).

In the past decade, extracellular vesicles (EVs) have been introduced as important modulators of angiogenesis. Recently, research on EVs has grown exponentially, highlighting their critical importance in vessel development in health and disease. Exosomes, subset of EVs with the diameter ranging from 30-150 nm and of endocytic origin, are now considered to be a key player in cell-to-cell communication. These virus-size particles, which have a lipid bilayer membrane, are constitutively released by fusion of multivesicular bodies (MVB) with the parent cell membrane. Exosomes have a complex biogenesis which is initiated by an endocytic event at the plasma membrane and is driven either by the four endosomal sorting complexes required for transport (ESCRT) or by the specific lipid composition of the endosomal membrane (Guo et al. 2017; Whiteside 2016). Since all types of cells in multicellular organisms secrete exosomes, their biological functions are manifold and depend on the cell of origin. For instance, it was reported that exosomes play a role in coagulation, inflammation, antigen presentation and immune regulation (Belting and Christianson 2015). Here, we discuss the intriguing and provocative role of exosomes in angiogenesis. This chapter will define the cargo of exosomes that impacts on their uptake by recipient cells such as endothelial cells (ECs), molecular pathways exosomes engage to reprogram ECs and their contribution to blood vessel formation. We will also consider recent insights into exosome potential as regulatory elements in angiogenesis.

CARGO OF EXOSOMES

Exosomes contain a variety of molecules such as proteins, nucleic acids (micro-RNAs, short and long noncoding RNAs), lipids and glycans. The exosome content reflects that of their parent cell and varies depending on the cell type (Boriachek et al. 2018). The protein content can be used for EV characterization, as recently defined by the International Society for Extracellular Vesicles (ISEV). Exosomes contain transmembrane or GPI-anchored proteins associated with the cell membrane or endosomes, such as tetraspanins (CD63, CD81) and integrins or cell/tissue specific proteins such as EPCAM in epithelial cell-derived exosomes or PECAM1 in EC-derived exosomes (Théry et al. 2019). It has been reported, that exosomes carry proteins, which are derived from the endocytic compartment, such as TSG101 and ALIX. In addition to those proteins, which are often used for EV characterization, exosomes carry cytokines and growth factors as functionally-active components of their cargo, such as TGFβ1/2, IFN-γ and VEGF-A (Théry et al. 2019). It has also been reported that exosomes from different sources carry a variety of pro-angiogenic proteins, e.g., VEGF, IGFBP3, IL-8 and LOXL2 (Kucharzewska et al. 2013; J. H. Yoon et al. 2014). Those proteins can either be bound to the surface of the exosomes or are carried inside the vesicles (Van Niel, D'Angelo, and Raposo 2018).

In the vesicles lumen, several known classes of RNA have been described (Van Balkom et al. 2015). *In vitro* experiments showed that some mRNAs present in the lumen of exosomes could be translated into proteins in target cells (Boriachek et al. 2018). Nearly all classes of exosomal RNAs can mediate angiogenesis. Numerous investigators demonstrated the pro-angiogenic effects of small exosomal mRNAs (Skog et al. 2012; Van Balkom et al. 2015; Tomohiro Umezu et al. 2014). More recently, Lang et al. showed similar angiogenic effects induced by long non-coding RNA (Lang et al. 2017).

Besides proteins and nucleic acids, exosomes also carry lipids, although the lipid composition is less representative of the parent cell than the protein exosome content (Boriachek et al. 2018). Compared to parental cells, exosomes are enriched in sphingomyelins but not in cholesterol and mainly carry ceramides or other sphingolipid components (Boriachek et al. 2018).

To summarize, exosomes carry a complex cargo, which could affect target cells in a variety of ways. The complexity of the cargo is strictly dependent on the packaging system in the parent cell and can be altered by environmental conditions, such as hypoxia (Kucharzewska et al. 2013) and various pharmacological treatments (Ribeiro et al. 2013). Even culture conditions can have drastic effects on the exosome production rate and the cargo, and optimization of culture conditions should be carefully considered in studies of exosomes (Ludwig et al. 2019), especially before using exosomes for cell-free treatments (Patel et al. 2017; Huber et al. 2008). Also, genetic modifications of the parent cells, e.g., transfections, are possible if a change in the exosome cargo/biological functions is desired. Umezu et al. reported that transfection of bone marrow stromal cells with anti-angiogenic miR-340 led to miR-340 expressing exosomes, which caused anti-angiogenic effects as a result of exosome/EC co-incubation assays (Tomohiro Umezu et al. 2017). The exosome cargo can also be directly modified by electroporation. Ma et al. used exosomes for miRNA delivery and electroporated mesenchymal stromal cell (MSC)-derived exosomes with miR-132 to enhance their pro-angiogenic effects. Remarkably, miR-132 increased the expression of the specific target genes in HUVECs and stimulated tube formation and *in vivo* angiogenesis capacity (T. Ma et al. 2018).

INTERACTION OF EXOSOMES WITH ENDOTHELIAL CELLS

Exosomes can interact with ECs in a variety of ways. The fastest way of interaction is through ligand/receptor signaling. Yoon et al. were able to demonstrate that certain pro-angiogenic effects of exosomes were based on surface interactions. The authors showed that activation of early growth response-1 (Egr-1) in ECs by tumor-derived exosomes (TEX) stimulated EC migration. Inhibiting the ERK1/2 and JNK signaling pathways inhibited these TEX-mediated effects (Y. J. Yoon et al. 2014). In addition to surface interactions, exosomes are rapidly internalized by ECs. Thus, confocal microscopy studies showed that labeled exosomes are internalized by ECs within 15 min of co-incubation and immediately after internalization, exosomes localize to the perinuclear region (Ludwig et al. 2018). This finding suggests that endocytosis might be the main uptake mechanism of exosomes by ECs. However, the use of pharmacological inhibitors for different internalization pathways indicated that ECs can also utilize other mechanisms to take-up and internalize exosomes, such as phagocytosis, micropinocytosis, or lipid raft-mediated internalization (Ludwig and

Whiteside 2018; Mulcahy, Pink, and Carter 2014). The cargo of the exosomes is hereby transferred to the recipient cells. Exosomes can either deliver proteins or nucleic acids. For instance, nasopharyngeal carcinoma-derived exosomes deliver ICAM-1 and CD44v5 to ECs and promote EC migration and tubulogenesis (Chan et al. 2015). Multiple myeloma-derived exosomes were found to contain miR-135b, which enhanced EC tube formation via the HIF-FIH signaling pathway (Tomohiro Umezu et al. 2014).

ROLES OF TUMOR-DERIVED EXOSOMES
IN ANGIOGENESIS

The role of exosomes in cancer has been studied extensively in the past years. Interestingly, patients with cancer show elevated levels of exosomes in the plasma compared to healthy individuals (Peinado et al. 2012), and similar observations have been made by comparing the exosome secretion of cultured tumor cells in comparison to 'normal' cells (Ludwig et al. 2018). These findings have led to an enormous interest in exploring the biological functions of exosomes in cancer and introduced a new, previously unrecognized, player to the tumor microenvironment (TME). To date, exosomes are considered to play a role in tumorigenesis, tumor growth, tumor immune escape, drug resistance and in metastasis (Guo et al. 2017). Besides that, there is evidence that tumor-derived exosomes (TEX) promote angiogenesis in the TME in various types of cancer (Ludwig and Whiteside 2018). In 2012, Skog et al. reported that TEX released by glioblastoma cells contain mRNA, miRNA and are enriched in angiogenic proteins, including VEGF and FGF. The authors described the internalization of TEX by ECs and functional reprogramming of ECs in response to TEX uptake. Thus, the study described a novel mechanism for delivering genetic information as well as proteins to recipient cells in the TME to promote angiogenesis (Skog et al. 2012). Follow-up studies by different groups showed similar results in other types of cancer, including breast cancer (Maji et al. 2016), renal cancer

(Grange et al. 2011), lung cancer (Hsu et al. 2017) and head and neck cancer (Ludwig et al. 2018).

Several different groups of investigators independently reported on interactions of TEX with ECs, which led to functional reprogramming of the recipient cells, stimulating proliferation, migration and tube formation (Ludwig et al. 2018; Mineo et al. 2012). Similar results were also seen *in vivo* or *ex vivo,* using angiogenesis models such as the matrigel plug models (Mineo et al. 2012), aortic ring assays (Zeng et al. 2018), xenograft models (Kucharzewska et al. 2013) or orthotopic tumor models (Ludwig et al. 2018). These published results could be explained by different underlying mechanisms. As described above, nucleic acids carried by TEX can be transported to EVs and translated into proteins. Wang et al. were able to show that AML-derived exosomes contained VEGF and VEGFR mRNA and that co-incubation studies induced VEGFR expression in HUVECs (Wang et al. 2018). Mineo et al. reported on activation of Src-signaling by TEX (Mineo et al. 2012). Additionally, miRNA has been shown to be a key player in several exosome-mediated biological effects including the promotion of angiogenesis (T Umezu et al. 2013). Several publications address miRNAs, which are carried by TEX and which directly affect ECs and stimulate angiogenesis in this way, as summarized in Table 1. It should be remembered, however that each miRNA can target many different targets, and that validation of the effects of miRNAs present in TEX on EC altered functions is necessary before it can be accepted and confirmed. Several, but not all reports quoted in Table 1 are accompanied by such a validation. The encapsulation of miRNAs in the vesicles protects them from enzymatic degradation en route and enhances their stability compared to miRNAs freely circulating in body fluids. It was also reported, that TEX metabolically reprogram ECs by enhancing glycolysis, which is essential for survival, migration, and growth of ECs (Wang et al. 2018).

Importantly, the content of exosome cargos and exosome-mediated pro-angiogenic effects are dependent on environmental factors. Hypoxia is a main regulator of tumor progression and angiogenesis. Kucharzewska et al. described how cancer cells adapt to hypoxic conditions by releasing exosomes, which stimulate the phenotypic modulation of ECs more effective compared to exosomes produced under normoxic conditions. The proteome and mRNA profiles of exosomes from normoxic or hypoxic conditions highly reflected the oxygenation status of the donor cell/tissue (Kucharzewska et al. 2013). This was also reported regarding the enrichment of miRNAs in TEX. Table 1 lists miRNAs, which are enriched in TEX produced in hypoxic conditions.

Table 1. TEX carry miRNAs, which can be transferred to ECs causing functional reprogramming to a pro-angiogenic phenotype[a]

Cellular origin of exosomes	miRNA	References
Normoxic conditions		
Breast cancer	miR-126a	(Sondike, Pisetsky, and Luzier 2017)
Colorectal cancer	miR-25-3p	(Zeng et al. 2018)
Gastric cancer	miR-130a	(Yang et al. 2018)
Hepatocellular carcinoma	miR-16-5p, -21-5p, -122-3p, -122-5p, -195-3p, -195-5p, -199a-3p, -199a-5p, -210, -221-3p, -222-3p, -222-5p, -223-3p, -224-5p	(Yukawa et al. 2018; Lin et al. 2018)
Lung cancer	miR-21, -23a, -142-3p, -210	(Hsu et al. 2017; Y. Liu et al. 2016; Cui et al. 2015; Lawson et al. 2018)
Ovarian cancer	miR-21	(Cappellesso et al. 2014)
Upregulation under hypoxic conditions		
Hepatocellular carcinoma	miR-155	(Matsuura et al. 2018)
Leukemia	miR-18b, -20a, -24, -106b, -130b, -185, -210, -224, -379, -652	(Tadokoro et al. 2013)
Lung cancer	miR-23a	(Hsu et al. 2017)
Multiple myeloma	miR-135b, -200c, -210, -223, -328, -335, -425	(Tomohiro Umezu et al. 2014)

[a]A variety of miRNAs were found in the cargo of TEX secreted by different cancer cells, which are reported to have direct effects on ECs. Under hypoxic conditions TEX were found to have altered miRNA profiles including miRNAs, which affect the EC behavior and stimulate blood vessel formation.

Interactions of exosomes and EVs with ECs were also utilized for anti-angiogenic cancer treatments. Umezu et al. isolated bone marrow stromal cells and transfected them with miR-340. MiR-340 was present in exosomes, which were isolated from those cells and mediated anti-angiogenic effects by the HGF/c-MET signaling pathway in HUVECs. This exosome-based treatment effectively inhibited multiple myeloma-induced angiogenesis *in vivo* (Tomohiro Umezu et al. 2017). Similar results were reported by Lu et al. (Lu et al. 2018).

The available literature supports the role of TEX in stimulating angiogenesis by reprogramming cells of various types in the TME. In this respect, silencing of exosome-mediated effects might be beneficial for cancer patients and might enhance the effects of current anti-angiogenic therapies. Follow-up studies are necessary to explore whether TEX can be used as biomarkers for tumor vascularization and response to anti-angiogenic therapy.

MEDIATION OF ANGIOGENESIS IN THE CARDIOVASCULAR SYSTEM BY EXOSOMES

The angiogenesis-promoting potential of exosomes has also been described in the cardiovascular system. After ischemia, the regeneration and tissue repair is dependent on the induction of angiogenesis. The role of exosomes derived from different cell types in the cardiovascular system was, therefore, investigated extensively in the recent years, highlighting exosome importance.

Beltrami et al. demonstrated that the pericardial fluid contains elevated numbers of exosomes. Compared to exosomes from the peripheral plasma, the exosomes derived from pericardial fluid were enriched in miRNAs and the authors showed, that exosomes originating from the pericardial fluid orchestrated vascular repair via miRNA transfer (Beltrami et al. 2017). Similar results were found after co-incubation of serum-derived exosomes from coronary blood of patients with myocardial ischemia. Exosomes from patients with myocardial ischemia enhanced EC proliferation, migration and tube formation compared to co-incubation with exosomes from healthy donors. It was found, that these exosomes are mainly derived from cardiomyocytes, but not from cardiac fibroblasts or ECs. *In vivo*, exosomes derived from ischemic cells significantly promoted blood flow recovery and neovascularization (H. Li et al. 2018).

Exosomes derived from cardiomyocytes were also proposed as a cell-free approach to induce angiogenesis in patients with myocardial infarction. Dougherty et al. isolated exosomes from human induced pluripotent stem cell-derived cardiomyocytes and described the pro-angiogenic effects on ECs including stimulated proliferation and migration and enhanced expression of growth factors after uptake of exosomes (Dougherty et al. 2018).

Similar to exosomes derived from tumor cells, exosomes from cells in the cardiovascular system are highly affected by environmental conditions. Specifically, hypoxia, which is the main driving factor for angiogenesis, can lead to the secretion of exosomes with an increased angiogenic potential. Hypoxic conditions alter the cargo of exosomes and up-regulate several exosomal miRNAs. It has been shown, that exosomes secreted by cardiosphere-derived cells under hypoxia have increased levels of pro-angiogenic miRNAs, such as miR-126, miR-130a and miR-210 (Namazi et al. 2018). This leads to the idea, that exosomes, which might be used as a pro-angiogenic cell-free treatment should be isolated from cells grown under hypoxia to increase exosome potency.

In conclusion, exosomes from cells of the cardiovascular system, especially cardiomyocytes, seem to play a protective role and are involved in the regeneration and repair of tissue after cardiovascular diseases. In addition to those autologous exosomes, studies currently in progress illustrate the benefits of exosomes derived from cultured cells and indicate their usefulness for regenerative therapies.

MESENCHYMAL STROMAL CELL-DERIVED EXOSOMES

Mesenchymal stromal cells (MSCs) have been widely used for tissue regeneration (K. Qi et al. 2018). Autologous or allogenic MSCs mediate a variety of beneficial effects in different diseases and in tissue repair. It has been suggested that exosomes derived from MSCs can be used as an alternative to cell-based regenerative therapies. This approach overcomes the problem of low cell survival after transplantation (Dougherty et al. 2018). The advantages of MSC-derived exosomes relate to their varied and biologically-active cargo, and the absence of immunogenicity. Thus, administration of cell-free exosomes in place of MSCs might be favorable for tissue regeneration.

MSC-derived exosomes were characterized in detail by Anderson et al. who used a proteomics approach to investigate the cargo of MSC-derived exosomes (Anderson, Johnathon D. Johansson et al. 2016). These investigators identified several putative paracrine effectors of angiogenesis. Among others, platelet derived growth factor (PDGF), epidermal growth factor (EGF), fibroblast derived growth factor (FGF) and the nuclear factor-kappa-B (NKκB) signaling pathway proteins were detected in the MSCs cargo (Anderson, Johnathon D. Johansson et al. 2016). Zou et al. compared the miRNA content of MSC-derived exosomes with their cells of origin and showed that more than 98% of miRNAs showed the same expression in both MSCs and EVs produced by these cells. The miRNAs in MSCs and in EVs targeted the same signaling pathways modulating angiogenesis, oxidative stress and inflammation. Despite those extensive similarities between MSCs and EVs, the GO analysis indicated that miRNAs in EVs could affect different target genes through the specific signaling pathways (Zou et al. 2018), suggesting a broader functional repertoire of EVs compared to MSCs. One crucial factor for working with MSC-derived exosomes is a consistent and optimized MSC culture, since the environmental conditions influence the cargo of released exosome as well as their purity. In this context, the procedure used for exosome isolation from MSC supernatants could influence their cargo. Patel et al. described the optimization of culture conditions for exosome production by MSCs. The protocol provides a guideline for increasing the exosome yield and maximizing their vascularization bioactivity (Patel et al. 2017). Another consideration for improving the effectiveness of MSC-derived exosomes might be the use of a 3D culture instead of 2D cultures as reported recently (Y. Zhang et al. 2017).

The regenerative effects of MSC-derived exosomes, especially the stimulation of angiogenesis, have been reported in different settings. First, the biological functions of MSC-derived exosomes show promising effects in repairing injured tissues, especially in healing of cutaneous wounds. Thus, MSC-derived exosomes were shown to promote collagen synthesis and angiogenesis, which are key components for a succesful cutaneous wound healing (J. Zhang et al. 2015). Similar studies have been made in a diabetic rat model. Wound healing in diabetic patients is often complicated and accompanied by numerous risks. In this setting, exosomes from MSCs were found to have beneficial effects when used as an additive to hydrogel/chitosan scaffolds. Besides enhancing angiogenesis, the exosomes promoted re-epithelization, deposition and remodeling of collagen and neuronal ingrowth, which led to the conclusion that MSC-derived exosomes offer a practical value in skin repair (Shi et al. 2017).

Benefits of treatments with MSC-derived exosomes were also reported in traumatic brain injury. It is well known, that the transplantation of MSCs improves functional recovery after the brain injury. Zhang et al. demonstrated that a systemic administration of MSC-derived exosomes as a cell-free therapy can lead to beneficial effects comparable to those of MSC transplantation. In an experimental traumatic brain injury rat model, exosomes from bone marrow-derived MSCs were injected into the tail vein and functional recovery was monitored. Exosomes promoted angiogenesis and neurogenesis and reduced neuroinflammation, leading to an improved functional recovery (Y. Zhang et al. 2017). Similar effects were seen after systemic treatment of stroke or spinal cord injury with exosomes in a rat model (Xin et al. 2013; Huang et al. 2017). Nakamura et al. investigated the role of MSC-derived exosomes in skeletal muscle regeneration and showed that exosomes can promote myogenesis and angiogenesis. However, the concentrations of muscle-repair-related cytokines was found to be low in MSC-derived exosomes. Instead, the authors found higher numbers of repair-related miRNAs, suggesting that the observed effects were miRNA mediated (Nakamura et al. 2015).

MSC-derived exosomes have also been used as a treatment in bone regeneration. The pro-angiogenic properties of MSC-derived exosomes were hereby used to enhance angiogenesis and osteogenesis under osteoporotic conditions (X. Qi et al. 2016). Besides beneficial effects on bone regeneration, the MSC-derived exosomes were reported to have preventative effects on osteonecrosis in a steroid-induced rat osteonecrosis model. In this model, exosomes promoted local angiogenesis and prevented bone loss (X. Liu et al. 2017).

The paracrine role of MSCs in tissue regeneration is of great current interest. Due to the biologically-active cargo of MSC-derived exosomes, their effects are similar to those mediated by treatments with MSCs. Despite the presence of MHC class I and II antigens on exosomes, they are not immunogenic and are not rejected in the allogeneic setting. Modifying the exosomes cargo by specially selected culture conditions or genetic alterations of MSCs is a promising approach for regenerative therapies. Hereby, the exosome cargo could be adapted to the patient-specific disease pattern to ensure the delivery of pro- or anti-angiogenic factors at the right time and in the right dose to mediate repair and promote healing.

ENDOTHELIAL CELL-DERIVED EXOSOMES

Additionally to the research in which ECs are used as the recipient cells in co-cultures with exosomes from different sources, exosomes derived from ECs have also been studied. Van Balkom et al. characterized EC-derived exosomes by using next generation sequencing (NGS). The analysis revealed the presence of all known non-coding RNA classes in exosomes, including miRNAs, small nucleolar RNAs, yRNAs, vault RNAs, 5p and 3p fragments of miRNAs, miRNA-like fragments and additionally fragments of messengerRNAs (mRNAs) and mitochondrial RNAs (mtRNAs). The study showed, that the distribution of small RNAs in exosomes showed a considerable overlap with the distribution in the producing cells (Van Balkom et al. 2015). The same group also demonstrated that exosomes, which are secreted by ECs promote EC migration and angiogenesis *in vitro* and *in vivo*. The exosomes derived from human microvascular endothelial cells are enriched in miR-214, suppress senescence and induce angiogenesis in human and mouse ECs (Van Balkom et al. 2017). Because of the close histologic location, ECs have a strong relationship with surrounding vascular smooth muscle cells (VSMCs). Qiu et al. provided important information about the communication between those two cell types by performing LC-MS/MS-based proteomics on exosomes from VSMCs. The analyses revealed a large number of proteins, which probably have potential regulatory functions in the VSMC-EC communication. These exosomes did not stimulate angiogenesis, suggesting that the major function of VSMC-derived exosomes is to maintain vessel homeostasis (Qiu et al. 2018).

It was also shown, that the exosome production by ECs and the protein cargo of exosomes can be altered under different growth conditions. Conditioning of ECs with ethanol increased the EV production and the ability of endothelial EVs to induce a pro-vascularization response. The EVs hereby reflect the status of the parent cells, since ethanol induces an angiogenic phenotype in ECs. Ethanol-conditioned EVs also show an increase of long non-coding RNA content, leading to the idea that alcohol consumption may activate endothelial EVs towards a pro-vascularization phenotype. This observation could have implications for alcohol-induced tumor angiogenesis (Lamichhane et al. 2017). The production of exosomes by ECs was also stimulated by using low-level laser irradiation (LLLI). Although EC survival is slightly reduced by LLLI at high power intensity, the cells express higher levels of vesiculation-related genes and the exosome biogenesis is increased by engaging the transcription factors promoting the Wnt signaling pathway (Bagheri et al. 2018). It was also reported, that exosomes-derived from ECs have the potential as biomarkers for patients with cerebrovascular disease, since those exosomes carry higher levels of atherosclerosis-promoting proteins (Goetzl et al. 2017).

Besides exosome isolation from mature ECs, endothelial progenitor cells (EPCs) were also used as a source of exosomes. During blood vessel formation, the recruitment of bone-marrow-derived and/or vascular-wall-resident EPCs occurs (Carmeliet and Jain 2011). It is known that EPCs include many subtypes of cells, some of which will differentiate into ECs and promote vascular growth. Other cell subtypes stimulate angiogenesis by paracrine mechanisms (Marçola and Rodrigues 2015). Exosomes have been described as an active component of the paracrine secretion of human EPCs. Li et al. reported that EPC-derived exosomes promoted the re-endothelialization after endothelial damage and therefore accelerated the vascular repair (X. Li et al. 2016). The same group and Zhang et al. also showed, that EPC-derived exosomes facilitate wound healing by stimulating EC function (X. Li, Jiang, and Zhao 2016; J. Zhang et al. 2016). The published data indicate that exosomes derived from EPCs might have protective effects for ECs injury under physiologic conditions and that the potency of those effects can be increased by exercise (C. Ma et al. 2018). Two independent groups reported that EPC-derived exosomes are enriched in miR-126, which can be delivered to ECs. The knockdown (KO) of miR-126 diminished exosome functions *in vitro* indicating the importance of miR-126 for the protective and pro-angiogenic effects of EPC-derived exosomes (Wu et al. 2018; C. Ma et al. 2018).

However, working with EPC-derived exosomes as a cell-free therapeutic approach can be difficult, since the cargo and functional behavior was reported to be highly dependent on environmental factors. Specifically, inflammatory stimuli, which are encountered in ischemic tissues, can impair EPC as well as exosome function and have to be addressed in future studies (Yue et al. 2017). The loss of function and decreased exosomes production were also reported after exposing EPCs to diabetic conditions (Hassanpour et al. 2018).

CONCLUSION

The increasing number of publications in angiogenesis-related exosome research strongly advocates the importance of exosomes in blood vessel formation. Further, exosomes are implicated in various physiological processes such as the cell-to-cell communication between ECs and VSMCs, in the EC protection/repair and in the repair of wounds and tissues. Today, many questions remain unanswered about the mechanisms mediating these various protective effects, and future studies are necessary to explore the role of exosomes in the vascular homeostasis. In addition to their beneficial effects, exosomes also drive malignant diseases by accelerating angiogenesis in the cancer tissue. TEX might serve as angiogenesis biomarkers in the future, and it is possible, that through their vessel-reprogrammming effects, TEX interfere with cancer therapies. If so, then silencing of TEX and specifically their pro-angiogenic and immunosuppressive effects might introduce novel approaches to current anti-angiogenic therapies.

The use of exosomes in lieu of MSCs as a cell-free therapy in regenerative medicine is a promising direction. The possibility of modifying the exosome cargo by culture conditions or transfection of the cells producing exosomes allows for the delivery of exosomes carrying a defined cargo to recipient cells. In this fashion, it is possible to produce exosomes that will exert either pro- or anti-angiogenic effects (Ribeiro et al. 2013).

Although there are many emerging possibilities for exosome utilization in therapy of various diseases, the exosome field is currently in urgent need for an universally accepted exosome isolation method and for the defined exosome nomenclature. At present, neither exosome classification nor a gold standard method for their isolation exist, which creates difficulties with comparisons of results and results interpretation. The nature of EVs, their yields and purity remain unclear in the published reports that are quoted in this chapter. Not all of the discussed results can be ascribed to a subset of small vesicles called exosomes and could be due to effects mediated by larger microvesicles. Considering the promise of EVs, as future therapeutic factors in angiogenesis, it seems important to determine whether all or only a single population of EVs have the ability to alter angiogenesis.

REFERENCES

Anderson, Johnathon D. Johansson, Henrik J., Calvin S. Graham, Mattias Vesterlund, Missy T. Pham, Matt S. Bramlett, Charles S. Montgomery, Elizabeth N. Mellema, Renee L. Bardini, Zelenia Contreras, et al. 2016. "Comprehensive Proteomic Analysis of Mesenchymal Stem Cell

Exosomes Reveals Modulation of Angiogenesis via Nuclear Factor-KappaB Signaling." *Stem Cells* 34 (3):601–13.

Bagheri, Hesam Saghaei, Monireh Mousavi, Aysa Rezabakhsh, Jafar Rezaie, Seyed Hossein Rasta, Alireza Nourazarian, Çigir Biray Avci, et al. 2018. "Low-Level Laser Irradiation at a High Power Intensity Increased Human Endothelial Cell Exosome Secretion via Wnt Signaling." *Lasers in Medical Science* 33 (5). Lasers in Medical Science:1131–45.

Balkom, Bas W. M. Van, Almut S. Eisele, D. Michiel Pegtel, Sander Bervoets, and Marianne C. Verhaar. 2015. "Quantitative and Qualitative Analysis of Small RNAs in Human Endothelial Cells and Exosomes Provides Insights into Localized RNA Processing, Degradation and Sorting." *Journal of Extracellular Vesicles* 4 (2015):1–14.

Balkom, Bas W. M. Van, Olivier G. De Jong, Michiel Smits, Jolanda Brummelman, Krista Den Ouden, Petra M. De Bree, Monique A. J. Van Eijndhoven, et al. 2017. "Endothelial Cells Require miR - 214 to Secrete Exosomes That Suppress Senescence and Induce Angiogenesis in Human and Mouse Endothelial Cells." *Blood* 121 (19):3997–4007.

Belting, M., and H. C. Christianson. 2015. "Role of Exosomes and Microvesicles in Hypoxia-Associated Tumour Development and Cardiovascular Disease." *Journal of Internal Medicine* 278 (3):251–63.

Beltrami, Cristina, Marie Besnier, Saran Shantikumar, Andrew I. U. Shearn, Cha Rajakaruna, Abas Laftah, Fausto Sessa, et al. 2017. "Human Pericardial Fluid Contains Exosomes Enriched with Cardiovascular-Expressed MicroRNAs and Promotes Therapeutic Angiogenesis." *Molecular Therapy* 25 (3):679–93.

Boriachek, Kseniia, Md Nazmul Islam, Andreas Möller, Carlos Salomon, Nam Trung Nguyen, Md Shahriar A. Hossain, Yusuke Yamauchi, and Muhammad J. A. Shiddiky. 2018. "Biological Functions and Current Advances in Isolation and Detection Strategies for Exosome Nanovesicles." *Small* 14 (6):1–21.

Cappellesso, Rocco, Andrea Tinazzi, Thomas Giurici, Francesca Simonato, Vincenza Guzzardo, Laura Ventura, Marika Crescenzi, Silvia Chiarelli, and Ambrogio Fassina. 2014. "Programmed Cell Death 4 and microRNA 21 Inverse Expression Is Maintained in Cells and Exosomes from Ovarian Serous Carcinoma Effusions." *Cancer Cytopathology* 122 (9):685–93.

Carmeliet, Peter. 2003. "Angiogenesis in Health and Disease." *Nature Medicine* 9 (6):653–60.

Carmeliet, Peter, and Rakesh K. Jain. 2000. "Angiogenesis in Cancer and Other Diseases." *Nature* 407 (6801):249–57.

Carmeliet, Peter, and Rakesh K. Jain. 2011. "Molecular Mechanisms and Clinical Applications of Angiogenesis." *Nature* 473 (7347):298–307.

Chan, Yuk Kit, Huoming Zhang, Pei Liu, Sai Wah Tsao, Maria Li Lung, Nai Ki Mak, Ricky Ngok-Shun Wong, and Patrick Ying Kit Yue. 2015. "Proteomic Analysis of Exosomes from Nasopharyngeal Carcinoma Cell Identifies Intercellular Transfer of Angiogenic Proteins." *International Journal of Cancer* 137 (8):1830–41.

Cui, H., B. Seubert, E. Stahl, H. Dietz, U. Reuning, L. Moreno-Leon, M. Ilie, et al. 2015. "Tissue Inhibitor of Metalloproteinases-1 Induces a pro-Tumourigenic Increase of miR-210 in Lung Adenocarcinoma Cells and Their Exosomes." *Oncogene* 34 (28):3640–50.

Döme, Balázs, Mary J. C. Hendrix, Sándor Paku, József Tóvári, and József Tímár. 2007. "Alternative Vascularization Mechanisms in Cancer: Pathology and Therapeutic Implications." *The American Journal of Pathology* 170 (1):1–15.

Dougherty, Julie A., Naresh Kumar, Mohammad Noor, Mark G. Angelos, Mohsin Khan, Chun-an Chen, and Mahmood Khan. 2018. "Extracellular Vesicles Released by Human Induced-Pluripotent Stem Cell-Derived Cardiomyocytes Promote Angiogenesis" *Front Physiol.* (2018 December). 9: 1794.

Goetzl, Edward J., J. B. Schwartz, Maja Mustapic, Iryna V. Lobach, Richard Daneman, Erin L. Abner, and Gregory A. Jicha. 2017. "Altered Cargo Proteins of Human Plasma Endothelial Cell–derived Exosomes in Atherosclerotic Cerebrovascular Disease." *FASEB Journal* 31 (8):3689–94.

Grange, Cristina, Marta Tapparo, Federica Collino, Loriana Vitillo, Christian Damasco, Maria Chiara Deregibus, Ciro Tetta, Benedetta

Bussolati, and Giovanni Camussi. 2011. "Microvesicles Released from Human Renal Cancer Stem Cells Stimulate Angiogenesis and Formation of Lung Premetastatic Niche." *Cancer Research* 71 (15):5346–56.

Guo, Wei, Yibo Gao, Ning Li, Fei Shao, Chunni Wang, Pan Wang, Zhenlin Yang, Renda Li, and Jie He. 2017. "Exosomes: New Players in Cancer (Review)." *Oncology Reports* 38 (2):665–75.

Hassanpour, Mehdi, Omid Cheraghi, Belal Brazvan, Amirataollah Hiradfar, Nasser Aghamohammadzadeh, Reza Rahbarghazi, and Mohammad Nouri. 2018. "Chronic Exposure of Human Endothelial Progenitor Cells to Diabetic Condition Abolished the Regulated Kinetics Activity of Exosomes." *Iranian Journal of Pharmaceutical Research* 17 (3):1068–80.

Hsu, Y. L., J. Y. Hung, W. A. Chang, Y. S. Lin, Y. C. Pan, P. H. Tsai, C. Y. Wu, and P. L. Kuo. 2017. "Hypoxic Lung Cancer-Secreted Exosomal miR-23a Increased Angiogenesis and Vascular Permeability by Targeting Prolyl Hydroxylase and Tight Junction Protein ZO-1." *Oncogene* 36:4929–42.

Huang, Jiang-Hu, Xiao-Ming Yin, Yang Xu, Chun-Cai Xu, Xi Lin, Fu-Biao Ye, Yong Cao, and Fei-Yue Lin. 2017. "Systemic Administration of Exosomes Released from Mesenchymal Stromal Cells Attenuates Apoptosis, Inflammation, and Promotes Angiogenesis after Spinal Cord Injury in Rats." *Journal of Neurotrauma* 34 (24):3388–96.

Huber, Veronica, Paola Filipazzi, Manuela Iero, Stefano Fais, and Licia Rivoltini. 2008. "More Insights into the Immunosuppressive Potential of Tumor Exosomes." *Journal of Translational Medicine* 6 (1):63.

Kucharzewska, Paulina, Helena C Christianson, Johanna E Welch, Katrin J Svensson, Erik Fredlund, Markus Ringnér, Matthias Mörgelin, Erika Bourseau-Guilmain, Johan Bengzon, and Mattias Belting. 2013. "Exosomes Reflect the Hypoxic Status of Glioma Cells and Mediate Hypoxia-Dependent Activation of Vascular Cells during Tumor Development." *Proceedings of the National Academy of Sciences of the United States of America* 110 (18):7312–17.

Lamichhane, T. N., C. A. Leung, L. Y. Douti, and S. M. Jay. 2017. "Ethanol Induces Enhanced Vascularization Bioactivity of Endothelial Cell-Derived Extracellular Vesicles via Regulation of MicroRNAs and Long Non-Coding RNAs." *Sci Rep* 7 (1):13794.

Lang, Hai Li, Guo Wen Hu, Bo Zhang, Wei Kuang, Yong Chen, Lei Wu, and Guo Hai Xu. 2017. "Glioma Cells Enhance Angiogenesis and Inhibit Endothelial Cell Apoptosis through the Release of Exosomes That Contain Long Non-Coding RNA CCAT2." *Oncology Reports* 38 (2):785–98.

Lawson, James, Christopher Dickman, Rebecca Towle, James Jabalee, Ariana Rani, and Cathie Garnis. 2018. "Extracellular Vesicle Secretion of miR-142-3p from Lung Adenocarcinoma Cells Induces Tumor Promoting Changes in the Stroma through Cell-Cell Communication." *Molecular Carcinogenesis*, 1–12.

Li, Hao, Yiteng Liao, Lei Gao, Tao Zhuang, Zheyong Huang, Hongming Zhu, and Junbo Ge. 2018. "Coronary Serum Exosomes Derived from Patients with Myocardial Ischemia Regulate Angiogenesis through the miR-939-Mediated Nitric Oxide Signaling Pathway." *Theranostics* 8 (8):2079–93.

Li, Xiaocong, Chunyuan Chen, Liming Wei, Qing Li, Xin Niu, Yanjun Xu, Yang Wang, and Jungong Zhao. 2016. "Exosomes Derived from Endothelial Progenitor Cells Attenuate Vascular Repair and Accelerate Reendothelialization by Enhancing Endothelial Function." *Cytotherapy* 18 (2):253–62.

Li, Xiaocong, Chunyu Jiang, and Jungong Zhao. 2016. "Human Endothelial Progenitor Cells-Derived Exosomes Accelerate Cutaneous Wound Healing in Diabetic Rats by Promoting Endothelial Function." *Journal of Diabetes and Its Complications* 30 (6):986–92.

Lin, Xue Jia, Jian Hong Fang, Xiao Jing Yang, Chong Zhang, Yunfei Yuan, Limin Zheng, and Shi Mei Zhuang. 2018. "Hepatocellular Carcinoma Cell-Secreted Exosomal MicroRNA-210 Promotes Angiogenesis In Vitro and In Vivo." *Molecular Therapy - Nucleic Acids* 11:243–52.

Liu, Xiaolin, Qing Li, Xin Niu, Bin Hu, Shengbao Chen, Wenqi Song, Jian Ding, Changqing Zhang, and Yang Wang. 2017. "Exosomes Secreted from Human-Induced Pluripotent Stem Cell-Derived Mesenchymal Stem Cells Prevent Osteonecrosis of the Femoral Head by Promoting Angiogenesis." *International Journal of Biological Sciences* 13 (2):232–44.

Liu, Yi, Fei Luo, Bairu Wang, Huiqiao Li, Yuan Xu, Xinlu Liu, Le Shi, et al. 2016. "STAT3-Regulated Exosomal miR-21 Promotes Angiogenesis and Is Involved in Neoplastic Processes of Transformed Human Bronchial Epithelial Cells." *Cancer Letters* 370 (1):125–35.

Lu, Juan, Qi Hui Liu, Fan Wang, Jia Jie Tan, Yue Qin Deng, Xiao Hong Peng, Xiong Liu, Bao Zhang, Xia Xu, and Xiang Ping Li. 2018. "Exosomal miR-9 Inhibits Angiogenesis by Targeting MDK and Regulating PDK/AKT Pathway in Nasopharyngeal Carcinoma." *Journal of Experimental and Clinical Cancer Research* 37 (1):1–12.

Ludwig, Nils, Beatrice M. Razzo, Saigopalakrishna S. Yerneni, and Theresa. L. Whiteside. 2019. "Optimization of Cell Culture Conditions for Exosome Isolation Using Mini-Size Exclusion Chromatography (Mini-SEC)." *Experimental Cell Research* 378 (2):149–57.

Ludwig, Nils, and Theresa L. Whiteside. 2018. "Potential Roles of Tumor-Derived Exosomes in Angiogenesis." *Expert Opinion on Therapeutic Targets* 22 (5):409–17.

Ludwig, Nils, Saigopalakrishna S. Yerneni, Beatrice M Razzo, and Theresa L. Whiteside. 2018. "Exosomes from HNSCC Promote Angiogenesis through Reprogramming of Endothelial Cells." *Molecular Cancer Research* 16 (11):1798–1808.

Ma, Chunlian, Jinju Wang, Hua Liu, Yanyu Chen, Xiaotang Ma, Shuzhen Chen, Yanfang Chen, Ji Bihl, and Yi Yang. 2018. "Moderate Exercise Enhances Endothelial Progenitor Cell Exosomes Release and Function." *Medicine and Science in Sports and Exercise* 50 (10):2024–32.

Ma, Teng, Yueqiu Chen, Yihuan Chen, Qingyou Meng, Jiacheng Sun, Lianbo Shao, Yunsheng Yu, et al. 2018. "MicroRNA-132, Delivered by Mesenchymal Stem Cell-Derived Exosomes, Promote Angiogenesis in Myocardial Infarction." *Stem Cells International* 2018:1–11.

Maji, Sayantan, Pankaj Chaudhary, Irina Akopova, Phung M Nguyen, Richard J Hare, Ignacy Gryczynski, and Jamboor K Vishwanatha. 2016.

"Exosomal Annexin A2 Promotes Angiogenesis and Breast Cancer Metastasis." *Molecular Cancer Research* 15 (1):93–105.

Marçola, Marina, and Camila Eleuterio Rodrigues. 2015. "Endothelial Progenitor Cells in Tumor Angiogenesis: Another Brick in the Wall." *Stem Cells International* 2015.

Matsuura, Yusuke, Hiroshi Wada, Hidetoshi Eguchi, Kunihito Gotoh, Shogo Kobayashi, Mitsuru Kinoshita, Masahiko Kubo, et al. 2018. "Exosomal miR-155 Derived from Hepatocellular Carcinoma Cells Under Hypoxia Promotes Angiogenesis in Endothelial Cells." *Digestive Diseases and Sciences* 3/2019.

Mineo, Marco, Susan H. Garfield, Simona Taverna, Anna Flugy, Giacomo De Leo, Riccardo Alessandro, and Elise C. Kohn. 2012. "Exosomes Released by K562 Chronic Myeloid Leukemia Cells Promote Angiogenesis in a Src-Dependent Fashion." *Angiogenesis* 15 (1):33–45.

Mulcahy, Laura Ann, Ryan Charles Pink, and David Raul Francisco Carter. 2014. "Routes and Mechanisms of Extracellular Vesicle Uptake." *Journal of Extracellular Vesicles* 3:1–14.

Nakamura, Yoshihiro, Shigeru Miyaki, Hiroyuki Ishitobi, Sho Matsuyama, Tomoyuki Nakasa, Naosuke Kamei, Takayuki Akimoto, Yukihito Higashi, and Mitsuo Ochi. 2015. "Mesenchymal-Stem-Cell-Derived Exosomes Accelerate Skeletal Muscle Regeneration." *FEBS Letters* 589 (11). Federation of European Biochemical Societies:1257–65.

Namazi, Helia, Elham Mohit, Iman Namazi, Sarah Rajabi, Azam Samadian, Ensiyeh Hajizadeh-Saffar, Nasser Aghdami, and Hossein Baharvand. 2018. "Exosomes Secreted by Hypoxic Cardiosphere-Derived Cells Enhance Tube Formation and Increase pro-Angiogenic miRNA." *Journal of Cellular Biochemistry* 119 (5):4150–60.

Niel, Guillaume Van, Gisela D'Angelo, and Graça Raposo. 2018. "Shedding Light on the Cell Biology of Extracellular Vesicles." *Nature Reviews Molecular Cell Biology* 19 (4). Nature Publishing Group:213–28.

Patel, Divya B., Kelsey M. Gray, Yasasvhinie Santharam, Tek N. Lamichhane, Kimberly M. Stroka, and Steven M. Jay. 2017. "Impact of Cell Culture Parameters on Production and Vascularization Bioactivity of Mesenchymal Stem Cell-Derived Extracellular Vesicles." *Bioengineering & Translational Medicine* 2 (2):170–79.

Peinado, Hector, M Alečković, S Lavotshkin, I Matei, B Costa-Silva, G Moreno-Bueno, M Hergueta-Redondo, et al. 2012. "Melanoma Exosomes Educate Bone Marrow Progenitor Cells toward a pro-Metastatic Phenotype through MET." *Nature Medicine* 18 (6):883–91.

Qi, Kai, Na Li, Zhenyu Zhang, and Gerry Melino. 2018. "Tissue Regeneration: The Crosstalk between Mesenchymal Stem Cells and Immune Response." *Cellular Immunology* 326:86–93.

Qi, Xin, Jieyuan Zhang, Hong Yuan, Zhengliang Xu, Qing Li, Xin Niu, Bin Hu, Yang Wang, and Xiaolin Li. 2016. "Exosomes Secreted by Human-Induced Pluripotent Stem Cell-Derived Mesenchymal Stem Cells Repair Critical-Sized Bone Defects through Enhanced Angiogenesis and Osteogenesis in Osteoporotic Rats." *International Journal of Biological Sciences* 12 (7):836–49.

Qiu, H., S. Shi, S. Wang, H. Peng, S. J. Ding, and L. Wang. 2018. "Proteomic Profiling Exosomes from Vascular Smooth Muscle Cell." *Proteomics Clin Appl* 12 (5):215–25.

Ribeiro, Mara Fernandes, Hongyan Zhu, Ronald W. Millard, and Guo-Chang Fan. 2013. "Exosomes Function in Pro- and Anti-Angiogenesis." *Current Angiogenesis* 2 (1):54–59.

Shi, Quan, Zhiyong Qian, Donghua Liu, Jie Sun, Xing Wang, Hongchen Liu, Juan Xu, and Ximin Guo. 2017. "GMSC-Derived Exosomes Combined with a Chitosan/silk Hydrogel Sponge Accelerates Wound Healing in a Diabetic Rat Skin Defect Model." *Frontiers in Physiology* 8 (NOV):1–16.

Skog, Johan, Tom Wurdinger, Sjoerd Van Rijn, Dimphna Meijer, Laura Gainche, Miguel Sena-esteves, William T. Curry Jr, Robert S. Carter, Anna M Krichevsky, and Xandra O Breakefield. 2012. "Glioblastoma Microvesicles Transport RNA and Protein That Promote Tumor Growth and Provide Diagnostic Biomarkers." *Nat Cell Biol.* 10 (12):1470–76.

Sondike, Stephen B., Emily M. Pisetsky, and Jessica L. Luzier. 2017. "Exosomes miR-126a Released from MDSC Induced by DOX Treatment Promotes Lung Metastasis." *Oncogene* 21 (1):133–36.

Tadokoro, Hiroko, Tomohiro Umezu, Kazuma Ohyashiki, Toshihiko Hirano, and Junko H. Ohyashiki. 2013. "Exosomes Derived from Hypoxic Leukemia Cells Enhance Tube Formation in Endothelial Cells." *Journal of Biological Chemistry* 288 (48):34343–51.

Théry, Clotilde, Kenneth W. Witwer, Elena Aikawa, Maria Jose Alcaraz, Johnathon D. Anderson, Ramaroson Andriantsitohaina, Anna Antoniou, et al. 2019. "Minimal Information for Studies of Extracellular Vesicles 2018 (MISEV2018): A Position Statement of the International Society for Extracellular Vesicles and Update of the MISEV2014 Guidelines." *Journal of Extracellular Vesicles* 8 (1). Taylor & Francis:1535750.

Umezu, T., K. Ohyashiki, M. Kuroda, and J. H. Ohyashiki. 2013. "Leukemia Cell to Endothelial Cell Communication via Exosomal miRNAs." *Oncogene* 32 (22):2747–55.

Umezu, Tomohiro, Satoshi Imanishi, Kenko Azuma, Chiaki Kobayashi, Seiichiro Yoshizawa, Kazuma Ohyashiki, and Junko H. Ohyashiki. 2017. "Replenishing Exosomes from Older Bone Marrow Stromal Cells with miR-340 Inhibits Myeloma-Related Angiogenesis." *Blood Advances* 1 (13):812–23.

Umezu, Tomohiro, Hiroko Tadokoro, Kenko Azuma, Seiichiro Yoshizawa, Kazuma Ohyashiki, and Junko H. Ohyashiki. 2014. "Exosomal miR-135b Shed from Hypoxic Multiple Myeloma Cells Enhances Angiogenesis by Targeting Factor-Inhibiting HIF-1." *Blood* 124 (25):3748–57.

Wang, Bin, Xiaoting Wang, Diyu Hou, Qian Huang, Weiwu Zhan, Canwei Chen, Jingru Liu, et al. 2018. "Exosomes Derived from Acute Myeloid Leukemia Cells Promote Chemoresistance by Enhancing Glycolysis-Mediated Vascular Remodeling." *Journal of Cellular Physiology*, 1–13.

Whiteside, Theresa L. 2016. "Tumor-Derived Exosomes and Their Role in Cancer Progression." *Advances in Clinical Chemistry* 74:103–41.

Wu, Xu, Zilong Liu, Lijuan Hu, Wenyu Gu, and Lei Zhu. 2018. "Exosomes Derived from Endothelial Progenitor Cells Ameliorate Acute Lung Injury by Transferring miR-126." *Experimental Cell Research* 370 (1):13–23.

Xin, Hongqi, Yi Li, Yisheng Cui, James J. Yang, Zheng Gang Zhang, and Michael Chopp. 2013. "Systemic Administration of Exosomes Released from Mesenchymal Stromal Cells Promote Functional Recovery and Neurovascular Plasticity after Stroke in Rats." *Journal of Cerebral Blood Flow and Metabolism* 33 (11):1711–15.

Yang, Haiou, Haiyang Zhang, Shaohua Ge, Tao Ning, Ming Bai, Jialu Li, Shuang Li, et al. 2018. "Exosome-Derived miR-130a Activates Angiogenesis in Gastric Cancer by Targeting C-MYB in Vascular Endothelial Cells." *Molecular Therapy* 26 (10):2466–75.

Yoon, Jong Hyuk, Jaeyoon Kim, Kyung Lock Kim, Do Hyeon Kim, Sun Ju Jung, Hyeongjoo Lee, Jaewang Ghim, et al. 2014. "Proteomic Analysis of Hypoxia-Induced U373MG Glioma Secretome Reveals Novel Hypoxia-Dependent Migration Factors." *Proteomics* 14 (12):1494–1502.

Yoon, Yae Jin, Dae Kyum Kim, Chang Min Yoon, Jaesung Park, Yoon Keun Kim, Tae Young Roh, and Yong Song Gho. 2014. "Egr-1 Activation by Cancer-Derived Extracellular Vesicles Promotes Endothelial Cell Migration via ERK1/2 and JNK Signaling Pathways." *PLoS ONE* 9 (12):1–18.

Yue, Yujia, Venkata Naga Srikanth Garikipati, Suresh Kumar Verma, David A. Goukassian, and Raj Kishore. 2017. "Interleukin-10 Deficiency Impairs Reparative Properties of Bone Marrow-Derived Endothelial Progenitor Cell Exosomes." *Tissue Engineering Part A* 23 (21–22):1241–50.

Yukawa, Hiroshi, Kaoru Suzuki, Keita Aoki, Tomoko Arimoto, Takao Yasui, Noritada Kaji, Tetsuya Ishikawa, Takahiro Ochiya, and Yoshinobu Baba. 2018. "Imaging of Angiogenesis of Human Umbilical Vein Endothelial Cells by Uptake of Exosomes Secreted from Hepatocellular Carcinoma Cells." *Scientific Reports* 8 (1):1–12.

Zeng, Zhicheng, Yuling Li, Yangjian Pan, Xiaoliang Lan, Fuyao Song, Jingbo Sun, Kun Zhou, et al. 2018. "Cancer-Derived Exosomal miR-25-3p Promotes Pre-Metastatic Niche Formation by Inducing Vascular Permeability and Angiogenesis." *Nature Communications* 9 (1):5395.

Zhang, Jieyuan, Chunyuan Chen, Bin Hu, Xin Niu, Xiaolin Liu, Guowei Zhang, Changqing Zhang, Qing Li, and Yang Wang. 2016. "Exosomes Derived from Human Endothelial Progenitor Cells Accelerate Cutaneous Wound Healing by Promoting Angiogenesis through Erk1/2 Signaling." *International Journal of Biological Sciences* 12 (12):1472–87.

Zhang, Jieyuan, Junjie Guan, Xin Niu, Guowen Hu, Shangchun Guo, Qing Li, Zongping Xie, Changqing Zhang, and Yang Wang. 2015. "Exosomes Released from Human Induced Pluripotent Stem Cells-Derived MSCs Facilitate Cutaneous Wound Healing by Promoting Collagen Synthesis and Angiogenesis." *Journal of Translational Medicine* 13 (1):1–14.

Zhang, Yanlu, Michael Chopp, Zheng Gang Zhang, Hongqi Xin, Changsheng Qu, Meser Ali, Ye Xiong, Henry Ford Hospital, Henry Ford Hospital, and Henry Ford Hospital. 2017. "Systemic Administration of Cell-Free Exosomes Generated by Human Bone Marrow Derived Mesenchymal Stem Cells Cultured under 2D and 3D Conditions Improves Functional Recovery in Rats after Traumatic Brain Injury." *Neurochem Int* 111:69–81.

Zou, Xiang Yu, Yongjiang Yu, Sihao Lin, Liang Zhong, Jie Sun, Guangyuan Zhang, and Yingjian Zhu. 2018. "Comprehensive miRNA Analysis of Human Umbilical Cord-Derived Mesenchymal Stromal Cells and Extracellular Vesicles." *Kidney and Blood Pressure Research* 43 (1):152–61.

Reviewed by

Rakesh K. Jain, Department of Radiation Oncology, Massachusetts General Hospital, Harvard Medical School, Boston, MA 02114.

In: Advances in Medicine and Biology   ISBN: 978-1-53616-181-6
Editor: Leon V. Berhardt   © 2019 Nova Science Publishers, Inc.

Chapter 2

UTERINE MYOCYTES: DEVELOPMENT, STRUCTURE AND FUNCTION

Olesia Moroz * *and Alexander Zholos*[†]

Department of Biophysics and Medical Informatics,
Taras Shevchenko National University of Kyiv, Kyiv, Ukraine

ABSTRACT

Uterine myocytes provide the vital function - maintenance of a certain tone and quiescence during gestation and strong repeated contractions in labour. To perform such a complicated pattern of activity myometrial smooth muscle cells possess a range of highly specialized structural features that originate in their embryogenesis. The function of uterine myocytes to a great extent depends on their ability to respond to the different regulatory signals (neural, hormonal, mechanical, local) and consequently alter their membrane potential and ion permeability – the basis of excitation-contraction coupling.

The problem of preterm labour as a result of incorrect regulation or improper functioning of myometrial cells is still outstanding. 10% of all newborns remain premature and require complicated and expensive care

---

* Corresponding Author's E-mail: olesia.moroz@gmail.com.
† Corresponding Author's E-mail: a.zholos@univ.net.ua.

during first months and later often show reduced quality of life. The rate of neonatal death is also higher in the premature group. On the other hand, weak irregular contraction of the myometrium known as uterine hypocontractility is the most common labour complication that could lead to foetal distress and/or postpartum haemorrhage.

Thus, investigation of uterine myocytes development, structure and function remains an important area of research. This review covers the latest findings on myometrial cells embryogenesis, their later phenotypic changes, structure and its relation to contractility, which is considered to be the main function of these cells. In particular, we focus on one the least understood aspects of the myometrium function, namely the myogenic mechanisms of myocyte contraction. A significant part of the chapter therefore outlines recent data about the role of mechanosensitive membrane ion channels in the regulation of contractility. Among these, special attention is given to Transient Receptor Potential channels (TRP), which are known to be significant part in the regulation of smooth muscle contraction. The mechanism of calcium related relaxation will also be outlined, as well as the role of calcium sensitisation in uterus contractility.

This review chapter would be interesting for students, scientists and clinicians.

Keywords: uterine myocytes, myometrium, excitation-contraction coupling, molecular mechanisms

ABBREVIATIONS

(E-C) coupling - excitation-contraction coupling

(E-M) coupling - electro-mechanical coupling

(P-M) coupling - pharmaco-mechanical coupling

$[Ca^{2+}]_i$ - intracellular free Ca^{2+} concentration

20α-HSD - 20α-hydroxysteroid dehydrogenase

APC - adenomatous polyposis coli

AR - androgen receptors

BK_{Ca} - large-conductance Ca^{2+}-activated K^+ channels

CaM - calmodulin

CAPs - contraction-associated proteins

CAV-1 - caveolin 1

CCCs - constitutively active cation channels

cKO - conditional knockout mice

COX - cyclooxygenase

CPI-17 - protein phosphatase 1 inhibitor with molecular mass of 17 kDa

cPLA$_2$ - cytoplasmic phospholipase A$_2$

CREB - cAMP-response element binding protein (transcriptional factor)

Cx43 - connexin 43

ECM - extracellular matrix

EET - epoxyeicosatrienoic acid

ER - oestrogen receptor

ERK - extracellular signal–regulated kinase

ESR1 - oestrogen receptor type α

ESR2 - oestrogen receptor 2

FAK - focal adhesion kinase

Forkhead - *Forkhead* box proteins

FP - PGF2α receptor

GPCRs - G protein-coupled receptors

HB-EGF - heparin-bound epidermal growth factor

hERG - human ether-a-go-go-related gene K$^+$ channels

HMGA2 - high mobility group A proteins

HNF3 - hepatocyte nuclear factor 3α

HOXA11 - homeobox A11 protein

HSPB - small heat shock protein B

IGF - insulin-like growth factor

IGF BPs - insulin-like growth factor binding proteins

IL-1β - interleukin 1β

ISL1 - islet 1, differentiation marker

K$_{ATP}$ - ATP- and nucleoside diphosphates-sensitive potassium channels

KCNQ - potassium voltage-gated channel subfamily Q

LRP1 - low-density lipoprotein receptor-related protein 1

MAPK1 - mitogen-activated protein kinase 1

MED12 - mediator complex subunit 12

MHC - myosin heavy chains

miRs - micro RNAs

MLC - myosin light chains

MYPT - myosin targeting subunit

NCX - Na^+/Ca^{2+} exchanger

NF-kB - nuclear factor -kB

NGF - nerve growth factor

OAG - 1-oleyl-2-acetylglycerol

OTR - oxytocin receptor

P4 - progesterone

PGHS - prostaglandin H synthase

PGR - progesterone receptors

PI3KPK/Akt - phosphoinositide-3-kinase–protein kinase B/Akt

PKC - protein kinase C

PMCA - plasma membrane Ca^{2+} ATPase

PRA - progesterone receptors type-A

PRB - progesterone receptors type-B

PTGS2 - prostaglandin-endoperoxide synthase 2

RhoA-kinase - Ras homolog gene family, member A

RKIP - rapidly accelerated fibrosarcoma kinase inhibitor protein

ROCs - receptor-operated channels

ROK - Rho-associated kinase

RUNX1 - Runt-related transcription factor 1

SACs - stretch-activated cation channels

SERCA - sarco/endoplasmic reticulum Ca^{2+}-ATPase

SK - small conductance Ca^{2+}-activated K^+ channel

SMAD - a family of proteins similar to the gene products of the *Drosophila* gene 'mothers against decapentaplegic' (Mad) and the *C. elegans* gene Sma

SMC - smooth muscle cells

SOCE - store-operated calcium entry

SOCs - store-operated channels

SR - sarcoplasmic reticulum

STIM (1-2) - stromal interaction molecule (1-2)

TEA - tetraethylammonium

TGF-β - transforming growth factor beta

TIMP2 - tissue inhibitor of metalloproteinases 2

TM - tropomyosin

TNF-β - tumor necrosis factor-beta

TREK-1 - stretch-activated, four-transmembrane domain, two-pore potassium channels (K2P)

TRP - transient receptor potential channels: TRPC (canonical), TRPM (melastatin), TRPV (vanilloid), TRPP (polycystin), TRPML (mucolipin), TRPA (ankyrin) channels

TTX - tetrodotoxin

uNK - uterine natural killer cells

WNT7 - wingless-type mmtv integration site family, member 7

ZEB - zinc finger E-box binding homeobox protein family

α2M - alpha-2-macroglobulin

1. INTRODUCTION

The problem of preterm labours, that make 10.6% of all deliveries according to the World Health Organization (Chawanpaiboon et al., 2019) leading to high costs for neonatal care, as well as a decrease in quality of life of these children, is central to the clinical obstetrics and related fields in Physiology, Biophysics, Biochemistry and Molecular Pharmacology today. Also, poor labour activity (dysfunctional labour or dystocia) is often leading to a foetalfoetal distress and/or postpartum bleeding. In the USA, it is the cause for more than one third of the Caesarean sections (Boyle et al., 2013). In order to address these problems, research models to study the mechanisms of myometrium function are being actively developed. Various regulatory mechanisms of the contractile response and the signalling cascade are investigated, while new approaches for correction of myometrium cell function and transduction of regulatory signals are being constantly introduced. All these research directions require in-depth knowledge of uterine myocytes development, structure and functioning.

2. UTERINE SMOOTH MUSCLE DEVELOPMENT

2.1. Embryogenesis

Human female reproductive system starts its formation from the 5th week of embryo development and after that undergoes growth and maturation during the late gestation and after birth. By the end of the 14th week epithelial and myometrium layers become separated and till the end of week 22nd uterine muscle layer becomes completely differentiated (Cunha et al., 2018). Since the 8th week of foetal development α-actin reactivity that indicates smooth muscle differentiation can be observed. By the 20th week this immunoreactivity is recorded for all parts of the urogenital canal and distinct myometrial development can be proved by cell staining (Robboy et al., 2017). Cramer and colleagues have defined three patterns of myometrium development from embryo stromal cells. By the 18th week intramural myometaplasia could be observed, when myocytes are differentiated on the outer wall of the uterus and are able to develop muscular tone. The second pattern - subserosal myometaplasia - could be detected in the second trimester of foetal development when mesenchymal cells are transformed into the muscular in subserosal zones. Inframucosal myometaplasia is the third one that can be detected at the same stage of foetal growth and is typical for endomyometrial junction zone of uteri corpus (Cramer et al., 2015).

Molecular mechanisms underlying uterine smooth muscle differentiation are poorly understood, but it is suggested that signals from uterine epithelium triggers the differentiation of myometrium from uterine mesenchyme (Chunha et al., 2017, 2018). Variation in uterine development between species, namely the shape of the uterus and amount of myometrial tissue, depends on the rate and location of Mullerian duct fusion, giving as a result two uteri in rodents, bicornuate in sheep and single uterus in primates and human (Spencer et al., 2005). Uterine mesenchymal differentiation in human reproductive tract is accompanied by the expression of particular mesenchymal differentiation markers and transcription factors, namely Homeobox A11 protein (HOXA11) and Runt-related transcription factor 1

(RUNX1), respectively. Besides, at the middle of gestation developing uterus demonstrates oestrogen receptor α (ESR1)-positive immunoreactivity, while being negative for progesterone receptors (PGR) and androgen receptors (AR) and differentiation marker ISL1 (Cunha et al., 2017). Foetal adrenal cortex is supposed to be a facilitator of perinatal burst of myometrium via increased placental oestrogen production (Cramer et al., 2015).

Wingless-Type MMTV Integration Site Family, Member 7A (WNT7) genes that encode special signalling molecules and act via catenins are found to be significant for proper uterine myocytes development. Arango et al. (2005) has demonstrated that in mice lacking β-catenin myometrial cells that have been initially differentiated normally were turned to adipocytes. β-catenin is supposed to suppress adipogenic transcription factors and/or down-regulate the expression of specific transcription factor MyoD, which is crucial for muscle development (Arango et al., 2005). Wnt4 gene in mice is found to contribute to the normal growth of the Müllerian duct and further cell migration and differentiation to form uterine layers. In case of WNT4 deficiency longitudinal and circular layers of mouse myometrium become indistinguishable (Prunskaite-Hyyryläinen et al., 2016). Another study has shown that mutation in Adenomatous polyposis coli (APC) gene in mesenchymal cells – a part of the WNT signalling system – led to myometrium disruption, poor cell development and consequently dramatically decreased chance of successful pregnancy (Wang et al., 2011).

Transforming growth factor beta (TGF-β) is another molecule that mediates foetal uterine myocytes development. Experiments using TGF-β receptor type 1 conditionally knock-out (*Tgfbr1* cKO) mice revealed the disruption of smooth muscle cells, their spatial disorganisation as well as impaired mesenchymal-epithelial interactions, which is probably vital for proper myometrium development (Li et al., 2011). Forkhead box proteins together with epatocyte nuclear factor 3-alpha (Forkhead/HNF3) transcription factor were also found to be essential for foetal myometrium development. Its expression was localized to the inner myometrium and when impaired the uterine muscle become hypertrophied and disorganized.

This transcription factor is involved in progesterone effects on myometrium in postnatal life (Bellessort et al., 2015).

Insulin-like growth factor I (IGI) and its binding proteins (IGFBPs), namely IGFBP-4 and 5, are expressed in perinatal period throughout myometrium and uterine stroma and are also involved in myometrial differentiation (Gu et al., 1999).

2.2. Phenotypical Changes through Life

During postnatal life the uterus and the myometrium in particular continue its development. This is supported by studies of gene expression (Hu et al., 2004). For rats and mice at postnatal days 3-5 uterine myocytes start their spatial orientation and form circular and then longitudinal layer of the myometrium. Since 15 day of postnatal life maturation of rodent uterus begins, which is characterised by the growth of all uterine layers (Brody andand Cunha, 1989).

In postnatal development uterine myocytes require oestrogenic action. Administration of oestrogen antagonists results in the impairment of inner myometrium thickness, continuity, organisation, and bundlinging and even indicating the change in oestrogen signalling as a possible cause in the development of uterine adenomyosis (Mehasseb et al., 2009). IGF is supposed to be a mediator of oestrogen proliferator effect on uterine tissues postnatally. Maximal expression of IGF-I was detected in adult rats during the proestrus phase of cycle (Carlsson and Billig, 1991). IGFs are present in human and animal myometrium during oestrous cycle and regulate myocytes proliferation (Shynlova et al., 2007).

For postnatal modification of myometrium TGF-β appears to be required. In Tgfbr1 conditional knockout (cKO) mice the set of smooth muscle markers was similar to that of control animals but of a different distribution. This was followed by improper synthesis of the main proteins from extracellular matrix (ECM) therefore altering the influence of basement membrane on myocyte phenotype (Gao et al., 2014). During gestation-associated changes in the TGF-β signalling pathways typical to uterine myocytes significant role belongs to SMAD proteins (a family of proteins similar to the gene products of the *Drosophila* gene 'mothers against decapentaplegic' (Mad) and the *C. elegans* gene Sma), which when absent in knock-out mice lead to the impairment of myometrium integrity, variable thickness and altered boundary between stroma and smooth muscle layer. Mutations in *Smad1/5/4-Amhr2 genes* result in impaired implantation and decidualisation and are considered as possible reasons of early pregnancy loss (Rodriguez et al., 2016).

Hormonal changes during puberty in a certain manner modify uterine smooth muscle. Ultrasound examination of uterine wall thickness in healthy prepubertal and pubertal girls revealed that uterine wall and volume become significantly increasing with puberty progression. Oestrogens produced by enlarged follicles under hypothalamo-pituitary activation are suggested as main stimulators of uterine growth (Hagen et al., 2015). It was found that oestradiol also stimulated β-adrenergic receptor expression in myocytes (Brauer andand Burnstocks, 1998). For mice uterus Wnt4 plays significant role in myometrium maturation and growth in puberty (Prunskaite-Hyyryläinen et al., 2016). There is evidence indicating the role of HOX proteins, a family of transcription factors, in pubertal uterine layers maturation (Colvin andand Abdullatif, 2013).

Pregnancy-induced changes are referred to as phenotypic modulation and comprise four stages: (i) early proliferative with manifestation of myocytes proliferation, (ii) intermediate synthetic, when cellular hypertrophy is followed by the increase of ECM components, (iii) contractile with arising expression of connexins, oxytocin and prostaglandin receptors, gamma-actin, calcium and large-conductance Ca^{2+}-activated K^+ channels (BK_{Ca}); (iv) highly active labour (Miftahof and Nam, 2011).

Hyperplasia of the myometrium in pregnancy is to a great extent mediated by IGF and their ligands. IGF1 gene expression is increased in the first half of pregnancy, when myocytes proliferate, indicating the contribution of IGF to molecular mechanism of female steroid hormones effect on myometrium alterations (Shynlova et al., 2007). IGF BPs that are needed for proper IGF transportation and their receptor binding (Robinson et al., 2000) appeared to participate in the progesterone pathway of myometrial differentiation during pregnancy. IGFBP1 *vs* IGF1 intersections trigger and then restrict hyperplasia of uterine smooth muscle cells. IGFBP6 is hypothesized to mediate progesterone (P4) hypertrophy in midpregnancy. IGFBP2 was found only in the circular layer of gravid horns indicating its role in a contractile phenotyping of myocytes in late pregnancy (Shynlova et al., 2007).

Myometrium quiescence during gestation is maintained mainly due to a specific phenotype pattern that is supported by a particular ratio of P4 to its receptors. As Renthal with co-authors reports, when functional activity of PGR is normal, P4 inhibits proinflammatory transcription factors (nuclear factor kB (NF-kB) for instance) and stimulates expression of transcriptional inhibitors from zinc finger E-box binding homeobox protein family (ZEBs), namely ZEB1, which block the promoters of contractile genes (Renthal et al., 2015). Near term a pregnant myometrium develops a phenomena of P4 functional withdrawal that is due to the diminishing of gestation-supportive role of P4 due to the alteration of expression ratio and function between different PGR at a high level of the circulating hormone (Mesiano, 2004). Among the mechanisms underlying this process a certain role belongs to the mediators of the so-called myometrial inflammation like NF-kB, expression of which is stimulated by interleukin 1β (IL-1β), and they both suppress the genes that are upregulated by PGR (Lee et al., 2012).

Phenotypic changes of uterine myocytes during pregnancy also include the alteration of ion channels expression and distribution on the cell membrane being the basis of the changed excitability. Firstly, it was found that in the late gestation excitability of smooth muscle cells (SMC) is increased due to the inhibition of K^+ fast-inactivating transient outward current together with the shift of expression of Ca^{2+}-activated channels and the decrease of Ca^{2+} sensitivity of contractile myofilaments (Wang et al., 1998).

Myosin phosphatase, one of the key enzymes in myocytes contractility, appears to be a part of pregnancy-associated modification. Expression of its major component myosin targeting subunit (MYPT), which directs a catalytic subunit to dephosphorylate myosin II light chains, can be part of the molecular transition between uterine quiescence and activation (Lartey et al., 2016).

Micro RNAs (miRs) are also found to be modulators of myometrium phenotype in late gestation. As Nothnick reports, the miR-200 family affects uterine contractility during pregnancy and labour by regulating ZEB1 and ZEB2 expression (Nothnick, 2016). Another study revealed that miR-200 family could contribute to contractile phenotype via increasing the expression of 20α-hydroxysteroid dehydrogenase (20α-HSD), a key P4-metabolizing enzyme (Williams et al., 2012).

During late pregnancy and near term uterine myocytes undergo the effects of proinflammatory molecules produced by macrophages and neutrophils that have penetrated uterine tissue. NF-kB increases expression of genes that facilitate contractility, including prostaglandin F2α receptors, cyclooxygenase-2, oxytocin receptors (OTR), gap junction protein connexin 43 (Renthal et al., 2015). Members of the small heat shock protein B (HSPB) family that function as molecular chaperones also play important roles in cell death regulation, cytoskeletal rearrangements, and immune system activation. These could be involved in proinflammatory mediators signalling pathways changing the myometrium phenotype to the active state (MacPhee and Miskiewicz, 2017).

During postpartum period the uterine involution takes place so the processes of cellular remodelling change in proliferation rate, and increased susceptibility to infections takes place. In this regard toll-like receptors are being shown to participate in the reduction of myometrial thickness and autolysis of myometrial cells (Alan and Liman, 2018).

2.3. Cell Regeneration and Transformation

Uterine myocytes at certain circumstances could be transformed into abnormal cells that are capable of uncontrolled division together with overproduction and accumulation of ECM and consequent uterine fibroids or leiomyomas formation (Stewart, 2001). Oestrogens and P4 are suggested to play a major role in stimulating their growth. This effect is being developed as suppression of myocytes differentiation and fibroblasts activation could be mediated by NF-kB pathway with the participation of focal adhesion kinase (FAK) activated by TGFβ and activin A. Fibroid is also followed with increasing of tissue inhibitors of metalloproteinase level by some growth factors that leads to extra production of ECM components (Ciebiera et al., 2018). TGF-β acting via SMAD proteins is found to increase myocytes migration to the nodules enriched with collagen and fibronectin (Commandeur et al., 2015; Feng et al., 2016). A role of Raf kinase inhibitor protein (RKIP) is also shown in the pathogenesis of leiomyomas (Janjusevic et al., 2016). Short form of High Mobility Group A proteins (HMGA2) appeared to induce myometrial cell transformation toward putative tumor-initiating leiomyoma cells (Mas et al., 2015). One more possible cause of fibroid formation is the mutation of mediator complex subunit 12 (MED12) that is detected in 70-80% of patients (Ciebiera et al., 2018). Because genetic instability is among the reasons of uterine fibroids, 45 miR expressions found to be deregulated under these conditions. MiR-498 could target *ZEB2* and miR-200 by affecting ZEBs, tissue inhibitor of metalloproteinase 2 (TIMP2) and some other proteins expression have been linked to cell proliferation and tumour growth (Nothnick, 2016).

Postnatal uterine muscle is not a stable tissue and is subject to regeneration by differentiation of myogenic precursor cells, as is also the case in skeletal muscle (Arango et al., 2005). Human and rodent uterus and the myometrium in particular possess special population of cells - side population (adult stem cells), that are supposed to participate in tissue remodelling during pregnancy and postpartum. Within the myometrium these cells maintain quiescent cell cycle status, can activate OTR formation and reconstitute a tissue when become stimulated (Maruyama et al., 2010). These progenitor cells are activated by oestrogens via ESR1 (Ono et al., 2007) and they use integrin α6 (CD49f) as the marker and for self-renewal (Krebsbach and Villa-Diaz, 2017).

3. UTERINE MYOCYTES STRUCTURE

3.1. Myometrium Tissue Organisation

Myometrium is formed by SMC (myocytes) that are embedded within connective tissue elements. Such an organisation facilitates the force transduction within the tissue. Bio potentials when generated by single myocyte are transmitted from cell to cell via gap junctions (Wynn and Jollie, 1989). In rat, uterine myometrium is formed by two distinct layers – the inner circular and the outer longitudinal, while human uterine myocytes are not so well spatially organized (Carstem and Miller, 1990). Human myometrium consist of plenty of trabeculae oriented in different directions, but mostly longitudinal near the outer side and more circular in the deeper zone (Carstem and Miller, 1990). Through last decades two models of human myometrium organisation are debated: (i) largely interwoven continuous bundle type, and (ii) circular fibres orientation in stratum sub-vasculature, while in the rest of the myometrium the fibres are disordered. Three dimensional modelling has suggested myometrium fasciculi to be a functional unit of the uterine smooth muscles as they form interacting network by dichotomising and intertwining with each other (Miftahof and Nam, 2011).

Myometrium contains from 37% to 57% of the extracellular space, which is filled by fibroblasts, mast cells, vessels, nerves, macrophages and neutrophils that become infiltrated near term of labour (Carstem and Miller, 1990). ECM around uterine myocytes is formed by collagen, elastin, glycoproteins and proteoglycans. Collagen fibres are inserted in thin spaces between cells and form fibrils within connective tissue septa. Their special feature for the myometrium is the direct contact with the myocyte cell membrane, while there is often no basal lamina. In the postpartum period a certain amount of collagen is resorbed (Bulbring et al., 1981). To form a tissue myocytes contact with each other via gap junctions, and physically interact with ECM by binding of their focal adhesion proteins to extracellular fibres. In pregnancy, the number of such contacts extremely increases and the myometrium changes the architecture of ECM from the fibrillar to the sponge-like (Miftahof and Nam, 2011). In the myometrium there is one more population of cells – telocytes, which are interstitial Cajal cells-like, which seem to influence the function of SMC (Hutchings et al., 2009). Having direct contacts with myocytes they are supposed to take part in excitation propagation within the tissue. Making close connection to stem cells, telocytes could be involved in regeneration and ECM organisation. During pregnancy their number in the myometrium decreases becoming the highest in the postpartum period. Telocytes are investigated as possible mechanosensors and sensors of steroid hormones level (Janas et al., 2018) and are discussed to be involved in leiomyoma pathogenesis (Varga et al., 2018).

Myometrial cells have special structures for the force transduction from contractile apparatus (myofilaments) to the cell membrane. Dense bands (plaques) (Miftahof and Nam, 2011) are one of them and are described as small spots containing α-actinin, localising near the inner side of the plasmalemma that are penetrated by actin filaments (Carstem and Miller, 1990). These bands are believed to be built from thin and intermediate filaments thus taking part in the regulation of cell shape and to certain extent in contraction. Dense bands also serve in cell-to-cell communication by formation of intermediate junctions (Bulbring et al., 1981). More resent data revealed that dense bands also contain vinculin, integrins, syndecans, proteins of focal adhesion, extracellular signalling, and serve to connect cytoskeleton with ECM and as mechano-sensors in tyrosine kinase signalling pathways. Up to 90% of myocytes volume is occupied by myofilaments, intermediate filaments, thin filaments and dense bodies that are cytoplasmic structures localized between filaments (Wynn and Jollie, 1989; Carstem and Miller, 1990).

3.2. Myometrial Cell Membrane – Caveolae, Lipid Rafts, Ion Channels and Receptors

The myometrium primarily consists of SMC, which generate contractions required for parturition. Myocyte contraction is generally initiated by transient increases in intracellular free Ca^{2+} concentration ($[Ca^{2+}]_i$) due to action potential discharge in response to membrane depolarisation that opens voltage-gated calcium, and sometimes sodium ion channels. Thus, like many other types of visceral smooth muscles, myometrium is an electrically excitable phasic smooth muscle tissue that can generate spontaneous and receptor agonist/hormone-evoked contractions (Bolton et al., 1999). Early simultaneous recordings of spontaneous electrical and contractile activities in pregnant human myometrium using the single sucrose-gap technique have already demonstrated complex shapes of action potentials (both spike-types and plateau), with which all contractions were well synchronised (Kawarabayashi et al., 1986). Oxytocin potentiated these contractions by enhancing the plateau phase, while spike-type action potentials in the presence of oxytocin developed a plateau.

The resting membrane potential of uterine myocytes varies between about -35 and -70 mV depending on species, muscle layer and gestational stage (Sanborn, 2000). Simultaneous recordings of membrane potential, $[Ca^{2+}]_i$ and contraction in human uterine strips obtained from the lower segment during caesarean delivery showed that the resting membrane potential became progressively less negative from 29 weeks (-70 mV) until term and labour (-55 mV), and that such changes were associated with a progressive increase in the frequency of spontaneous contractions (Parkington et al., 1999). Some forms of slow changes in the membrane potential termed the "slow waves" have also been recorded in the myometrium, which by themselves do not seem sufficient to initiate contractions, although these may help triggering action potential discharge (Tribe, 2001).

Since any change in membrane potential is caused by ion currents flowing through the plasmalemma it is reasonable to suggest that some finely tuned and coordinated changes in expression and function of various ion channels underlie such intricate changes in the resting membrane potential and action potential shape and frequency of discharge as seen in the myometrium depending on uterus functional state (e.g., non-pregnant *vs* pregnant, pregnancy stage etc.). In addition, electrogenic exchangers and ion pumps, such as the Na^+/K^+ ATPase, can also be expected to contribute to alterations in excitability and excitation process. In turn, changes in ion conductance of the plasmalemma may be myogenic, voltage- and/or calcium-dependent, or caused by activation of cell-surface receptors, second messengers, as well as stretch.

Hyper-K^+ external solution invariable causes membrane depolarisation and contraction of the myometrium, while these are strongly suppressed under extracellular Ca^{2+}-free conditions. These findings implicate voltage-gated Ca^{2+} channels and K^+ channels as the major effector channels. Conversely, smooth muscle relaxation is caused by membrane hyperpolarisation that reduces the frequency of action potential discharge and in any case causes decrease in the probability of L-type Ca^{2+} channels to be in an open state, thus reducing voltage-dependent Ca^{2+} influx and associated with it muscle tension. L-type Ca^{2+} channel blockers such as nifedipine or nitrendipine are known to attenuate or even abolish myometrial contractions (reviewed by Tribe, 2001). Other types of ion channels such as chloride-permeable and non-selective cation channels (e.g., receptor- and store-operated channels) play important, but not yet fully understood roles in myometrial electrogenesis.

Better understanding of the function of membrane proteins requires the knowledge of signalling microdomains and their structural counterparts, such as caveolae and lipid rafts, which are believed to play important, yet still incompletely understood role in the myometrium. Ultrastructure studies have shown that surface flask-shaped vesicles of 50-80 nm in diameter, the so-called caveolae, densely populate the plasma membrane of uterine myocytes contributing to about 30% of cell surface (Cole and Garfield, 1989). These structures containing marker proteins – caveolins – participate in signal transduction processes and are believed to orchestrate the dynamics of myometrial intracellular signalling. Caveolin-1 (CAV-1) expression at both mRNA and protein level remains unaltered in human pregnancy (Taggart et al., 2007). Especially characteristic for phasic smooth muscles, the caveolae are in close juxtaposition with peripheral sarcoplasmic reticulum (SR), where intracellular calcium is stored. Caveolin can interact with heterotrimeric G proteins. Furthermore, the actin cytoskeleton is also a part of a caveolar complex, which participates in the regulation of myometrial BK_{Ca} channel activity. In both non-pregnant and late-pregnant uterus BK_{Ca} channel proteins colocalize with caveolin-1 and caveolin-2 (but not caveolin-3), as well as with both alpha- and gamma-actin. Disruption of these caveolar complexes by cholesterol depletion or by cytochalasin D caused increase in iberiotoxin-sensitive K^+ current (Brainard et al., 2005). Oestrogens reduces the number of caveolae and caveolin expression in the myometrium (Turi et al., 2001).

Apart from caveolae, which may be regarded a type of lipid raft stabilised by caveolins, plasmalemma microdomains enriched in sphingolipids and cholesterol referred to as lipid rafts are under investigation for their role in regulating uterine calcium homeostasis and contractility. Notably, oxytocin activity was shown to be dependent on lipid rafts as the binding affinity of OTR was reduced by disruption of rafts (Klein et al., 1995). Furthermore, cholesterol extraction and raft disruption greatly potentiates spontaneous $[Ca^{2+}]_i$ rises and contractions of rat and human myometrium, while the opposite happens at elevated cholesterol content (Noble et al., 2006). These effects are, at least in part, mediated by the corresponding changes in the activity of BK_{Ca} channels, which provide, via a negative feedback, the inhibitory control of myometrial contractility.

As already mentioned, among the diverse molecular pathways for calcium entry, on which myometrial contraction ultimately depends, high-threshold voltage-gated L-type Ca^{2+} channels are predominantly expressed in the myometrium. As pregnancy progresses, spike-type potentials may acquire a prolonged plateau phase, for maintenance of which activation of Cl^- channels is important (Parkington and Coleman, 1990; Sanborn, 2000). Interestingly, plateau potentials at about -30 mV and lasting 30-40 s (Parkington et al., 1999) would provide an optimum for the so-called "window current" flowing via L-type Ca^{2+} channels resulting in a significant sustained Ca^{2+} entry for the contractile response. In other cell types, such currents arise due to the overlapping steady-state activation and inactivation curves of L-type Ca^{2+} channels, but its relevance to myometrial physiology remains to be established. In this connection it is worth mentioning that Ca^{2+}-activated Cl^- currents have been described in pregnant rat myometrium.

Inhibitors of these channels were shown to inhibit oxytocin-induced $[Ca^{2+}]_i$ transients and associated contractions, but not high-K^+-induced contractions (Jones et al., 2004).

In many smooth muscles the voltage-dependent calcium current is not homogenous and it is possible to distinguish low- and high-threshold currents by using appropriate voltage pulse protocols and holding potentials (Bolton et al., 1999). mRNAs and proteins encoding two T-type Ca^{2+} channels subunits (Ca_v 3.1 and Ca_v 3.2) were detected in the longitudinal muscle layer of pregnant rat myometrium. Blockers of these channels (mibefradil, NNC 55-0396 and nickel) reduced amplitude and frequency of spontaneous $[Ca^{2+}]_i$ transients and contractions suggesting a key role for T-type Ca^{2+} channels in the regulation of the frequency of spontaneous phasic contractions (Lee et al., 2009). Electrophysiological studies also showed two types of voltage-activated Ca^{2+} currents similar to the T- and L-type currents that exist in freshly dispersed, pregnant human uterine myocytes. Interestingly, magnesium (8 mmol/L) reduced T-type currents by about 70% but did not reduce L-type currents, while nifedipine blocked the L-type currents but not T-type currents (Young et al., 1993). Thus, T-type Ca^{2+} channels may potentially be involved in the initiation of action potential discharge in the myometrium, but their functional significance is not fully appreciated.

Ohkubo and co-workers (Ohkubo et al., 2005) have quantified mRNA expression of the two types of calcium channels in longitudinal and circular muscle layers of rat myometrium during pregnancy using a comparative kinetic RT/PCR method and concluded that during pregnancy the expression levels of L-type channels changed dynamically, and this may contribute directly to the regulation of cell excitability. In longitudinal smooth muscle the change in alpha1H mRNA encoding T-type Ca^{2+} channels was similar to that of L-type channels during gestation, but the alpha1G mRNA, also encoding T-type Ca^{2+} channels, changed significantly only at term (day 22). In the circular layer, alpha1G mRNA significantly increased only on day 15 and at term.

The importance of voltage-gated Ca^{2+} channels for myometrial contraction is highlighted by clinical usefulness of Ca^{2+} channel blockers, which are one class of tocolytics. Drugs such as nifedipine and diltiazem (L-type channel blockers) and mibefradil (T-type channel blocker) reduce and then abolish the contractility of preterm and term myometrium, although they are less efficient during labour (Arrowsmith et al., 2010).

It should be noted that other drugs may also display Ca^{2+} channel antagonistic properties. Among them, the non-steroidal anti-inflammatory drug nimesulide is a potentially useful tocolytic. It is a selective inhibitor of cyclooxygenase-2, which causes relaxation of spontaneously contracting human myometrium *in vivo*. Part of this effect has been attributed to the inhibition of T- and L-type Ca^{2+} channels, as found in freshly dispersed human term-pregnant myometrial SMC (Knock and Aaronson, 1999).

In contrast to the generally accepted prominent roles of voltage-gated Ca^{2+} channels in smooth muscle function, voltage-dependent sodium channels show restricted expression depending on smooth muscle type, species, and sometimes even age. Thus, we have previously described typical voltage-gated sodium current highly sensitive to tetrodotoxin (TTX) (IC_{50}=4.5 nM) in myocytes freshly isolated from rat ileum (Smirnov et al., 1992). Other studies have reported sodium currents with different TTX sensitivities, voltage dependence and kinetics in several other types of visceral smooth muscles including human myometrium (Bolton et al., 1999). Based on these variable properties it is likely that voltage-dependent Na^+ channels of smooth muscles are a heterogeneous group.

In cultured human uterine myocytes voltage dependence of Na^+ channels is such that they should be largely inactivated at physiological resting potentials (Young and Herndon-Smith, 1991). In single uterine muscle cells isolated from nonpregnant, pregnant and postpartum rat uteri both fast TTX-sensitive (IC_{50}=27 nM) and slow dihydropyridine-sensitive L-type Ca^{2+} channels are coexpressed (Ohya and Sperelakis, 1989; Martin et al., 1990; Yoshino et al., 1997). Moreover, Yoshino et al. (1997) have shown that the ratio of densities of peak I_{Na}/I_{Ca} changes from 0.5 in the nonpregnant state to 1.6 at term. They thus suggested that the larger contribution of I_{Na} of faster kinetics allows more frequent repetitive spike discharges facilitating simultaneous excitation of the parturient uterus. Furthermore, it is suggested that oestrogen-enhanced genomic influences may regulate increased densities of Na^+ and Ca^{2+} channels as pregnancy progresses. Inoue and Sperelakis also investigated gestational changes in the properties of Na^+ and Ca^{2+} channels in rat myometrial smooth muscle cells. Voltage-dependence of these channels and density of Ca^{2+} currents remained unaltered during gestation, while current density of fast Na^+ channels increased markedly due to an increase in fraction of myocytes expressing these channels (Inoue and Sperelakis, 1991). These findings again highlight importance of Na^+ channels in the spread of smooth muscle excitation. In the uteri of nonpregnant rats expression of several voltage-gated Na^+ channel alpha-subunits (Scn2a1, Scn3a, Scn5a, and Scn8a), as well as all beta-subunits have been detected, while functional tests revealed TTX-sensitive rhythmic contractions evoked by veratridine, a Na^+ channel activator (Seda et al., 2007).

Expression of the so-called contraction-associated proteins (CAPs) including connexin 43 (Cx43), OTR, prostaglandin-endoperoxide synthase 2 (PTGS2), prostaglandin H synthase (PGHS), PGF2α receptor (FP), and CAV-1 is the key factor determining gestational-dependent changes in myometrial contractility. These will be discussed in detail later in this review. However, during pregnancy the myometrium remains relatively quiescent, and here we shall focus our attention on another group of plasmalemmal ion channels that generate outward hyperpolarising currents thus controlling smooth muscle excitability and excitation process itself, namely K^+ channels. As an interesting parallel, vascular smooth muscles are generally non-excitable, largely due to high density and early activation of various types of K^+ channels present in vascular myocytes. In their recent elegant experiments Borysova and colleagues showed a dramatic change from normally restricted non-conducted vasoconstriction induced in rat mesenteric resistance arteries by local high-K^+ solution application to intercellular Ca^{2+} waves and vasoconstriction that spread along an entire artery segment 3 mm long in the presence of K^+ channel blocker tetraethylammonium (TEA) and L-type Ca^{2+} channel agonist BayK 8644. This was due to (artificial in this case) action potential generation spreading along gap junctions (Borysova et al., 2018). These observations illustrate the important notion that it is the balance of depolarising and hyperpolarising ion currents together with electrical syncytium function that determines the level of smooth muscle excitability, spread of excitation and contractility.

Different types of K^+ channels are expressed in uterine myocytes, but their precise roles and relative contributions to myometrium electrical activity remains to be fully understood. Some of these channels are open at the resting membrane potential, while others are predominantly activated by membrane depolarisation (K_V channels), $[Ca^{2+}]_i$ rise (BK_{Ca} channels) and other stimuli. ATP- and nucleoside diphosphates-sensitive potassium channels (K_{ATP}) open in response to reduced intracellular ATP concentration thus providing an important link between cellular metabolism and electrogenesis.

BK_{Ca} channels are some of the most ubiquitously distributed in various smooth muscles K^+ channels. These channels are activated in synergy by membrane depolarisation and $[Ca^{2+}]_i$ rise. Thus, they are ideally suited for membrane repolarisation following opening of voltage-gated Ca^{2+} channels during action potential discharge. Even at rest, these channels can open in response to highly localised $[Ca^{2+}]_i$ increases close to the plasmalemma, which occur due to sporadic activation of clusters of ryanodine receptors located on the (SR, known as "Ca^{2+} sparks." The resulting coordinated openings of up to 100 BK_{Ca} channels produce spontaneous transient outward currents (STOCs) (Bolton et al., 1999). Such functional coupling of ryanodine receptors to BK_{Ca} channels causes vasodilation of arterial smooth muscle since Ca^{2+} sparks have little direct impact on global $[Ca^{2+}]_i$, which regulates contraction, but provide sufficiently large local $[Ca^{2+}]_i$ elevations for BK_{Ca} activation, thus causing membrane hyperpolarisation and deactivation of voltage-dependent Ca^{2+} channels (Nelson et al., 1995; Wellman et al., 2002). However, it should be noted that Ca^{2+} sparks and corresponding STOCs are generally lacking in rat myometrium, which may be important for the generation of strong and long-lasting contractions of the uterus required in labour (Burdyga et al., 2007).

In gastrointestinal smooth muscles, global transient $[Ca^{2+}]_i$ rises occur in consequence to Ca^{2+}-induced Ca^{2+} release, which activate massive late outward Ca^{2+}-dependent K^+ current via BK_{Ca} channels (Zholos et al., 1991, 1992). However, in contrast to these and many other types of visceral and vascular smooth muscles whereby such roles of BK_{Ca} channels are well characterised, less is known about physiological settings for BK_{Ca} channel activation in the myometrium. Interestingly, BK_{Ca} channels somehow lose their Ca^{2+} sensitivity in the pregnant human and mouse myometrium during gestation and especially at the onset of labour (Khan et al., 1997; Benkusky et al., 2000) – one likely mechanism for modulation of myometrial excitability during pregnancy. Both decreased expression of BK_{Ca} alpha subunit at labour onset and increased proportion of its splice variants have been proposed to play a major role in decreasing voltage- and Ca^{2+} sensitivity of BK_{Ca} channels at the time of labour (Curley et al., 2004). In addition, BK_{Ca} beta-subunit expression, which regulates BK_{Ca} channel current kinetics and increases its apparent calcium and voltage sensitivity, is also reduced at term (Gao et al., 2009). Along these lines, downregulation of BK_{Ca} beta-subunit expression may also contribute to the pathophysiological preterm labour (Matharoo-Ball et al., 2003).

Several excellent reviews have been published summarising expression and properties of the predominant BK_{Ca} and other types of K^+ channels in pregnant and non-pregnant myometrium (Khan et al., 2001; Brainard et al., 2005, 2007; Greenwood and Tribe, 2014; Lorca et al., 2014). Here we thus focus on several recent studies and only several types of K^+ channels other than already discussed BK_{Ca} channels. Apart from these large-conductance channels, a specific isoform of small conductance Ca^{2+}-activated K^+ (SK) channel, SK3, is expressed in mouse myometrium whereby it promotes feedback regulation of myometrial Ca^{2+} and hence relaxation of the uterus (Brown et al., 2007). Apamin, the specific SK channel blocker, was shown to inhibit NO-induced relaxation of human non-pregnant myometrium (Modzelewska et al., 2003).

Stretch-activated, four-transmembrane domain, two-pore potassium channels (K2P), TREK-1, are expressed in human myometrium contributing to uterine quiescence during gestation, while their dysfunction may be involved in spontaneous pre-term birth (Buxton et al., 2010). Human ether-a-go-go-related gene (hERG) K^+ channels also perform similar function – their activity suppresses contraction amplitude and duration before labour, thereby facilitating quiescence, but closer to term the expression of beta-inhibitory subunits becomes markedly enhanced contributing to increased duration of uterine action potentials and contraction. Interestingly, this mechanism fails in obesity, thus contributing to weak contractions necessitating caesarean delivery (Parkington et al., 2014). Several different Kv channel alpha subunit proteins were found to be expressed in the myometrium from both non-pregnant and term-pregnant mice. Kv channel blocker 4-AP caused contractions of non-pregnant myometrium, but not in pregnant mice, and one of these proteins, namely Kv4.3, was found to disappear in term-pregnant myometrium (Smith et al., 2007).

Along with enzymes and several important second messenger signalling systems, such as cAMP/PKA, DAG/PKC, PLC/IP$_3$, ion channels represent very important group of effectors regulated by G protein-coupled receptors (GPCRs). This link in uterine myocytes is currently under intense investigation. There is rapidly accumulating evidence proving that ion channel regulation by GPCRs plays critically important roles in the modulation of myometrial excitability during pregnancy and at the onset of labour. Such GPCR-mediated regulation of ion channel activity and hence excitation of the myometrium is exemplified by oxytocin-induced inhibition of the Na$^+$-activated K$^+$ channel, Slo2.1 (Ferreira et al., 2019), modulatory effects of the K$_{ATP}$ channel openers (composed of SUR2B and Kir6.2 in murine myometrium) on spontaneous contraction, oxytocin- and PGF2α-induced contractions (Hong et al., 2016), functional coupling of β3-adrenoceptors and BK$_{Ca}$ channels in human myometrium (Doheny et al., 2005), activation of BK$_{Ca}$ channels by membrane-associated cGMP kinase contributing to uterine quiescence in pregnancy (Zhou et al., 2000), as well as by immunomodulatory interacting proteins such as alpha-2-macroglobulin (α2M) and its receptor, low-density lipoprotein receptor-related protein 1 (LRP1) in human myometrium (Wakle-Prabagaran et al., 2016). The latter mechanism is of special relevance to the below described mechanisms of regulation of uterine contractility by proimflammatory local factors (see Section 3.3).

Despite the wealth of knowledge about the above discussed myometrial plasma membrane ion channels and receptors, clinical usefulness of drugs, both the uterotonics and tocolytics, many of which actually target these transmembrane proteins directly or indirectly, remains rather disappointing (Arrowsmith et al., 2010). It is thus not surprising that much focus of the research on ion channels expressed in the myometrium is currently shifting towards the most recently discovered superfamily of Transient Receptor Potential (TRP) cation-selective channels (Venkatachalam and Montell, 2007). These channels when activated perform three major cellular functions: (i) many TRPs can admit Ca^{2+} directly as their Ca^{2+} permeability and density of plasmalemmal expression may be sufficiently high for inducing $[Ca^{2+}]_i$ rise sufficient for myocyte contraction; (ii) their activation invariable causes membrane depolarisation, which in cells expressing voltage-gated Ca^{2+} channels such as uterine myocytes, engages this additional powerful Ca^{2+} entry mechanism; and (iii) TRP channels are also expressed in intracellular organelles, in which case they function as intracellular Ca^{2+} release channels. Moreover, TRP channels can interact with many different proteins, e.g., they can form signalling complexes with BK_{Ca} channels and ryanodine receptors, and this way they are described as molecular conductors of a diverse orchestra. Importantly, many TRPs, particularly those of the "canonical" TRPC subfamily, are receptor-operated channels gated by activated G-proteins and second messengers such as DAG following GPCR stimulation, while other TRPs function as unique cellular sensors for a large array of physical (mechanical forces, temperature) and chemical (changes in pH, lipid environment etc) factors. Polymodal activation of these channels means that they function as signal integrators enabling cells to sense changes in their local environment and adequately respond to them. Lack of such responses, on the other hand, may be the primarily reason for cell dysfunction, and hence TRP channels are believed to offer many novel treatment opportunities. With the ever growing number of selective and potent TRP pharmacological modulators paralleled by our better understanding of specific roles of these channels in uterine smooth muscles it is reasonable expectation that such novel approaches to problems in pregnancy will soon be developed. However, caution is of course needed as the same TRP subtypes are expressed in different cell types, most notably in the CNS, and significant side effects may be an issue, but the same concern goes for the more "traditional" pharmacological modulators, for example blockers of L-type Ca^{2+} channels.

Based on their structural homology, 28 mammalian TRPs are classified into six subfamilies: TRPC (canonical), TRPM (melastatin), TRPV (vanilloid), TRPP (polycystin), TRPML (mucolipin), and TRPA (ankyrin) channels. Specific roles of TRP channels have been most extensively characterised in vascular smooth muscles (Guibert et al., 2011; Yue et al., 2015). Based on the analysis of signal transduction pathways, gating mechanisms and functional roles of cation currents mediated by vascular TRPs, Albert and Large have proposed their classification as constitutively active cation channels (CCCs), receptor-operated channels (ROCs), store-operated channels (SOCs) and stretch-activated cation channels (SACs) (Albert and Large, 2006). Such in-depth knowledge of myometrial TRPs is yet to be achieved, but described below examples show that at least one TRP isoform in each of these categories exists in uterine myocytes as well. Therefore, presumably this classification can be useful for the myometrium as well. It should also be noted that due to polymodal activation, as well as the existence of heterotetrameric channels composed of different TRP isoforms, the same TRP subtype can actually fulfil the criteria for more than one in the above listed functional subgroups.

Early studies have characterised the expression and putative roles of multiple TRPC isoforms (TRPC1/C3/C4/C6/C7) in store-operated calcium entry (SOCE) in pregnant human uterus (Dalrymple et al., 2002; Yang et al., 2002). In rat myometrium, TRPC1, TRPC2, TRPC4-C7 are expressed, while TRPC4 is particularly abundant (Babich et al., 2004). With the emergence of two other gene families, stromal interaction molecule (STIM) (1-2) and Orai (1-3) that mediate SOCE the major focus in this area of research has now shifted to these proteins. Indeed, all five genes have been shown to be expressed in pregnant human myometrium with Orai2 being the most abundant protein (Chin-Smith et al., 2014). Expression of STIM1-2/Orai1-3 did not change during labour onset, but interestingly Orai1 expression was upregulated by pro-inflammatory cytokine IL-1β.

Mechanical stretch of the uterus in pregnancy modulates myometrial growth, hyperplasia and hypertrophy, as well as contractility via alteration in calcium homeostasis. Increased expression of TRPC3 and TRPC4 in response to stretch was shown in human uterine myocytes (Dalrymple et al., 2007). IL-1β, a cytokine implicated in labour, specifically up-regulates TRPC3 expression in human myometrium (Dalrymple et al., 2004). TRPC3 has been implicated in labour initiation as its expression is enhanced in preterm labour, while TRPC3 knockout significantly delays inflammation-induced preterm labour in mice (Jing et al., 2018). TRPC3 overexpression was also found in lipopolysaccharide-induced preterm delivery mice model (Zheng et al., 2016).

Spontaneous myometrial contractions are altered with gestational age as their frequency and amplitude increase toward the end of gestation to initiate labour. TRPC4 and TRPC5 channels have been found to be involved in stretch-induced spontaneous uterine contractions of pregnant rat, thus these channels have been suggested as putative targets for treatment of premature labour (Chung et al., 2014). Multifunctional roles of TRPC4 channel are well illustrated by the observation that in human myometrial cells selective knockdown of endogenous TRPC4 using shRNAs specifically attenuates GPCR (oxytocin)-stimulated, but not thapsigargin- or 1-oleyl-2-acetylglycerol (OAG) -stimulated extracellular Ca^{2+}-dependent increases in $[Ca^{2+}]_i$ (Ulloa et al., 2009).

TRPs of other subfamilies are also expressed in female reproductive organs and placenta (Dörr and Fecher-Trost, 2011). Notably, their expression is not restricted to myocytes as these are found in nerve endings innervating reproductive organs (TRPV1) and oestrogens enhances responses to painful stimuli of the uterine cervix; in decidualized cells of the uterus and in uterine natural killer (uNK) cells (TRPV1); in ciliated epithelial cells of reproductive organs (TRPV4); in the luminal and grandular epithelium of the uterus (TRPV6). Low levels of TRPM4 and TRPM7 expression were detected in the uterus, and their functions remain to be established (Dörr and Fecher-Trost, 2011). Among these channels, TRPV4 might represent an especially interesting therapeutic target to address preterm labour. First, TRPV4 gene and protein expression is increased with gestation. Second, TRPV4-mediated Ca^{2+} entry and contractility were increased in uterine myocytes in pregnant *vs* nonpregnant rats. Third, oxytocin-induced myometrial contraction was reduced by TRPV4 inhibition and in mice with global deletion of TRPV4 (Ying et al., 2015).

Thus, various TRP subtypes may serve as important signal transduction elements of stretch, proinflammatory mediators and receptor agonist mediated signalling for the initiation of parturition.

3.3. Cell Contacts - Gap Junction, Role of Connexins, Signal Propagation

The excitation spreading from cell to cell and hence contraction coordination in the myometrium is possible due to the places of cell-to-cell contacts, which have high conductance. In the 1970th such structures were identified and referred to as gap junctions or nexuses and their presence in the myometrium was detected only for pregnant or parturient tissue (Garfield et al., 1978). Then it was found that gap junctions are formed by connexins, proteins that are arranged in hexamers, equal in both adjacent cells, and create a channel permeable for small metabolites, second messengers and ions (Doualla-Bell et al., 1995). Now a separate connexin family is described, but in gap junction formation connexin 43 (Cx43) is mostly involved, although there are data indicating that for human myometrium at term Cx43, 45 and 40 are included in the junctions (Kilarski et al., 2001). Their high adhesion rate could be explained by the presence of three inviolable cysteine residues in the two extracellular loops. Gap junctions could be formed by connexins of different types that defines their ion selectivity and permeability to cAMP, NAD or inositol polyphosphates (Evans and Martin, 2002). Cx43-containing gap junctions are necessary for labour beginning indicating that the transport of Cx43 to the myocyte membrane is more critical that the synthesis of this protein itself. Such a transport requires actin filaments that transfer specifically translated 6 N-terminally truncated isoforms of Cx43 to the plasma membrane (Nadeem et al., 2017).

Early studies identified that gap junction formation is sensitive to transcription and translation inhibitors and their size is regulated by prostaglandins, but not oxytocin (Garfield et al., 1980; MacKenzie and Garfield, 1985). Gap junction expression on myocytes is promoted by oestrogens and suppressed by P4. In rats Cx43 expression during pregnancy remains low because of inhibitory effect of P4, but before both term and preterm labour increases due to elevated oestradiol/P4 ratio that become higher as the luteosis progresses. In human myometrium Cx43 expression control is provided due to myocyte responsiveness to oestrogens and P4 influences and is based on switching of P4 effect from a transcriptional repressor to a transcriptional activator of Cx43 mediated via progesterone receptors type B (PRB) and progesterone receptors type A (PRA), respectively. Inhibitory effect of P4 on connexins expression is partially mediated by transcriptional co-repressor p54NRB (Kidder and Winterhager, 2015). Thus, gap junction formation and myocytes coupling at the onset of labour are mediated via PR type A (Nadeem et al., 2017). That is why maintaining of the uterus quiescence during pregnancy by P4 among other factors is provided by the reduction of gap junction formation due to inhibition of protein biosynthesis.

Gap junction density is upregulated by mechanical stretch in response to growing uterus (Wynn and Jollie, 1989). For bovine and rat uterus a difference in expression and regulation between longitudinal and circular layers of the myometrium exists with a prevalence of these contacts in the first one (Doualla-Bell et al., 1995). In experiments with knock-out mice Cx43 is found to mediate the onset of labour, because in case of its underexpression and reduced quantity of gap junctions 82% of animals had a delay of labour (Doring et al., 2006).

3.4. Intracellular Calcium Stores – Organisation and Function

As already mentioned, a distinguishable feature of phasic smooth muscles is closely associated with caveolae a network of subplasmalemmal SR tubes and sacs, a calcium storage site (Bolton et al., 1999). The gap between these structures is within 100 nm or so, but may be even less than 20 nm. Another SR compartment consisting of interconnecting tubules and cisternae is located deeper in the myocyte, throughout the cytoplasm, and especially around the Goldgi apparatus and nucleus. There are connections between these parts of the SR. Mitochondria are also increasingly implicated in Ca^{2+} uptake and storage. SR relative volume has been estimated at about 6%, increasing throughout pregnancy (Noble et al., 2006). Parallel upregulation of sarco/endoplasmic reticulum Ca^{2+}-ATPase (SERCA) ensures that larger amounts of calcium can be taken up by the SR.

Both IP_3 and ryanodine receptors are present on the SR membrane, but unlike other visceral smooth muscles Ca^{2+} sparks sporadically generated by clusters of ryanodine receptors do not seem to occur in uterine myocytes (Burdyga et al., 2007). In ileal myocytes, we have found predominant localisation of IP_3 type 1 receptor in the superficial SR, while ryanodine receptors are located in the deep SR (Gordienko and Zholos, 2004), but in uterine myocytes the precise locality of these Ca^{2+}-releasing channels is not known.

It is well known that most of the calcium entering smooth muscle myocytes is rapidly buffered and sequestered in various storage sites, with buffering ratio of more than 200, even before it reaches the contractile machinery. Superficial SR is particularly important in this calcium sequestration. This means that Ca^{2+} entry should be amplified by Ca^{2+}-induced Ca^{2+} release for full contraction to develop, and in most cases this is the function of the deep cytosolic SR (Young and Zhang, 2004). Between these zones the superficial calcium buffer barrier exists.

There is evidence that the SR plays an important role in the regulation of myometrial excitability and contractility. In particular, Ca^{2+} store depletion by the SR Ca^{2+} pump inhibition with thapsigargin or cyclopiazonic acid enhances $[Ca^{2+}]_i$ and myometrial contractions, and these effects are more pronounced in labour (Tribe et al., 2000; Noble et al., 2014). The above discussed STIM (1-2) and Orai (1-3) proteins, with likely contributions of some TRPC channels, in particular TRPC1 (Murtazina et al., 2011), underlie SOCE under these conditions, and SOCE seems to be the main pathway for

$[Ca^{2+}]_i$ increases and myometrial contractions, when the SR becomes Ca^{2+}-depleted (Chin-Smith et al., 2014).

All three subtypes of IP_3 receptors were found in both non-pregnant and pregnant human myometrium, expressed at different ratios in individual patients (Morgan et al., 1996). IP_3 receptors play central role in Ca^{2+} mobilisation by receptor agonists acting at $G_{q/11}$ coupled receptors, which activate phospholipase $C\beta$. In the context of the myometrium, such agonists are primarily oxytocin and PGE_2. The exact role of IP_3 receptors, as well as ryanodine receptors, in non-pregnant and pregnant myometrium remains largely debatable.

In addition to the SR, significant calcium storage capacity in myometrial SMC has been found to be provided by mitochondria (Marchi and Pinton, 2014). Kosterin and colleagues showed that in uterine myocytes Ca^{2+} accumulative capacity of mitochondria by more than 10 times exceeds that of sarcolemmal vesicles, which is accompanied by higher affinity of mitochondrial transports system for ionized Ca^{2+}. Based on their kinetic analysis, they suggested that mitochondria, while uptaking Ca^{2+} from the cytoplasm up to the level of 10^{-6}-10^{-7} M, promotes the functioning of plasmalemmal Ca^{2+} pump that contributes to myocyte relaxation (Kosterin et al., 1985). More recent data obtained by this group indicate that mitochondrial calcium uniporter is involved in NO-mediated Ca^{2+} uptake being involved in the inhibitory mechanisms of the regulation of uterine contractility (Danylovych et al., 2015). Reduced expression of this mitochondrial calcium uniporter was found in myometrium with gestation that can be the result of downregulation of transcriptional factor CREB in late pregnancy and labour providing higher cytosolic Ca^{2+} level needed for effective contractions. On the other hand, the appropriate ratio between activating MICU1 and the inhibitory MCUb could contribute to uterine myocyte hypertrophy (Vishnyakova et al., 2019). Another kind of Ca^{2+} transport system has been described for mitochondrial inner membrane, where the Ca^{2+}/H^+-exchanger is found to provide Ca^{2+} transport from organelle matrix to the myoplasm (Kolomiiets et al., 2014b) and found to be independent from NO but activated by calixarenes (Kolomiiets et al., 2014a). Thus, its functional role remains to be elucidated.

3.5. Myofilaments and Role of Cytoskeleton in Myofilament Organisation

Thick (myosin) filaments, or myofilaments, have a tail and head part formed by myosin heavy chains (MHC) and the head is covered with one regulatory (MLC20) and one essential (MCL17) myosin light chains (Taggart and Morgan, 2007). For rat uterus two MHC isoforms are reported, but there are data about MHC200 and MHC204, which are formed after alternative splicing in inserted and non-inserted modifications (Wynn and Jollie, 1989; Kelley and Adelstein, 1994). After these early studies it was questioned whether or not they form homo- or heterodimers in the cell. However, now for myocardial cells as well as for smooth muscles homodimers ($\alpha\alpha$ and $\beta\beta$) are described (Kopylova et al., 2016), where the contractility and filaments arrangement is related to exactly homodimeric organisation of myosin (Rovner et al., 2002). For humans the tissue specificity is confirmed and a difference in isoforms could be the cause of distinct contractile properties of different smooth muscles (Wynn and Jollie, 1989).

Contraction in myocyte is developed when cytosolic level of ionized calcium increases and MLC kinase becomes active. Phosphorylation of MLC enables the cross-bridges formation between actin filaments and myosin heads, Mg-ATPase activity of which provides the energy of conformational changes and tension development or shortening (Wray, 2007; Arrowsmith et al., 2014). The ATPase activity of myosin in uterine myocytes could be modulated and calix[4]arenes, which are cup-like polyphenolic compounds, are the example. They can penetrate the myocyte membrane, lower the contractility by up to 50% and change the structure of myofilaments (Labyntseva et al., 2016)

At term uterine myocytes have their contractile apparatus upregulated and possible pathways of such changes include reduced activation of MLC kinase, increased co-operative activation of actomyosin interactions at a given level of MLC20 phosphorylation or the altered expression of MLC20. Based on analysis of expression the last is supposed to be the most likely explanation (Taggart and Morgan, 2007).

In uterine myocytes thin filaments contain actin that is present in its two isoforms - α and γ, in nearly equal amounts. Gamma isoform increases its expression toward term (Wynn and Jollie, 1989; Taggart and Morgan, 2007).

Myometrial cells thin filaments contain also tropomyosin (TM), which is supposed to facilitate Mg-ATPase activity of myosin. TM forms helical structures near actin strands. In uterine myocytes the molar ratio of TM and monomer actin is 1:6 to 1:7 (Wynn and Jollie, 1989; Somara et al., 2005).

Thin filaments could be involved in the regulation of contractile function of myocytes via caldesmon that is a protein being able to bind in a Ca^{2+}-dependent manner with calmodulin (CaM) and Ca^{2+}-independently with actin and terminates contraction. Moreover, phosphorylation of caldesmon by Ca^{2+}/CaM dependant protein kinase blocks its inhibitory effect on myosin Mg-ATPase (Wynn and Jollie, 1989). Such properties of caldesmon, expression of which is elevated during gestation, make it one of the mechanisms contributing to the maintenance of uterine quiescence (Taggart and Morgan, 2007).

Cytoskeleton in uterine myocytes is formed by intermediate filaments and microtubules, which contain cytokeratin and tubulin, respectively. By passing through cytosolic dense bodies and membrane dense plaques these filaments support the cell shape and also take part in readying it for the labour contractile effort (Taggart and Morgan, 2007). FAK-Src system as one of the dense plaques proteins were shown to be activated by the mechanical stretch that results in extracellular signal–regulated kinase (ERK) activation and further MLC phosphatase suppression and contractile force enhancement (Li et al., 2007). According to Morgan the effect of mechanical stretch because of growing foetus/es could trigger the onset of labour due to disinhibition of the contractile filaments by phosphorylated caldesmon as the result of focal adhesion proteins activation and further ERK effect. It is suggested that increased cortical stiffness of myocytes that is observed in late pregnancy, when the uterine wall becomes distended, will cause the larger tension stimulation of focal adhesion proteins that will result in facilitation of contractility (Morgan, 2014).

4. FUNCTION OF UTERINE MYOCYTES

4.1. Excitation-Contraction and Pharmaco-Mechanical Coupling

In visceral smooth muscles, changes in membrane potential normally precede changes in tension, although contraction or relaxation of existing tone may occur independently of change in membrane potential (Bolton et al., 1999). As already discussed, action potential discharge rather than slow waves initiate myocyte contraction in the myometrium. The complex sequence of events taking place between action potential generation and muscle contraction is referred to as excitation-contraction (E-C) coupling. The key step in this process is the rise in $[Ca^{2+}]_i$, as this is needed to enable interaction between the two main proteins of the contractile apparatus, actin and myosin. Uterine cells contain various ion channels (although primarily these are L-type Ca^{2+} channels and possibly receptor-operated TRP channels) mediating Ca^{2+} entry during membrane depolarisation, as well as Ca^{2+} releasing channels, such as IP_3 and ryanodine receptors. Some TRP channels are likely to be present in intracellular organelles being thus able to sustain Ca^{2+} release, but this needs to be experimentally verified. Ca^{2+} then binds to CaM, which regulates smooth muscle contraction. Ca^{2+}-bound CaM interacts with MLC kinase, causing it to phosphorylate MLC at S19 or Y18 (Kuo and Ehrlich, 2015). The phosphorylated MLC by establishing cross-bridges with actin finally produces contractile force. Finally, both Ca^{2+} and Ca^{2+}-CaM complexes also bind to and regulate numerous other proteins in myocytes, including ion channels, protein kinases and transcription factors thus exerting short- and long-term effects. Elevated $[Ca^{2+}]_i$ stimulates effectors that reduce cytosolic Ca^{2+}, such as SERCA pump in the SR, plasma membrane Ca^{2+} ATPase (PMCA) and the Na^+/Ca^{2+} exchanger (NCX). Ca^{2+} then dissociates from CaM, which ultimately terminates contraction.

A wider definition of E-C coupling in sooth muscles apart from the above described electro-mechanical (E-M) coupling also includes the so-called pharmaco-mechanical (P-M) coupling, in which process contraction is not primarily associated with changes in membrane potential (Bolton et al., 1999). Molecular mechanisms of P-M coupling include activation of

GPCR leading to Ca^{2+} release from the store and a number of other mechanisms, by which smooth muscle tone can be modulated by G-proteins, protein kinases and phosphatases via phosphorylation/dephosphorylation reactions, and possibly other signalling molecules such as cAMP, cGMP and NO. Whether or not P-M coupling strictly confirms to the definition of E-C coupling can be debated, but clearly under physiological conditions these two forms (E-M and P-M) of E-C coupling co-exist and often act in synergy to produce maximally effective mechanism for force generation in smooth muscles.

The well-established scenario whereby myofilaments can generate stronger contraction at a constant, or nearly constant $[Ca^{2+}]_i$ is referred to as the phenomenon of Ca^{2+} sensitisation. It is based on those molecular mechanisms that reduce MLC phosphatase activity. Presently two such pathways are well established: Ras homolog gene family, member A (RhoA-kinase), whereby GTP-RhoA is an active substance and DAG-protein kinase C (PKC), which acts via protein phosphatase 1 inhibitor with molecular mass of 17 kDa (CPI-17) (Kuo and Ehrlich, 2015). PKC phosphorylates CPI-17 to prevent MLC phosphatase activity. When active molecules bind to the myosin binding subunit of MLC phosphatase the enzyme activity become reduced so the force of contraction increases. Rho-associated kinase (ROK) activated by GTP-RhoA and CPI-17 are found to be elevated near term thus taking part in labour contractility (Taggart and Morgan, 2007). Recently soluble MYPT1-unbound form of catalytic protein phosphatase 1cβ was found to contribute to MLC dephosphorylation and relaxation of smooth muscle as an alternative pathway (Chang et al., 2018).

We have previously showed that both Rho kinase and PKC are involved in vascular smooth muscle myofilament calcium sensitisation in arteries from diabetic rats (Kizub et al., 2010), while in resistance arteries myogenic response to increased intravascular pressure involves potentiation of the RhoA pathway (Cole and Welsh, 2011). Such potential triggers of Ca^{2+} sensitisation would be of importance in uterine (patho)physiology, but to our knowledge these have not been elucidated in the myometrium.

4.2. Mechanisms of Myometrial Excitation

It logically follows from the previous section that our better understanding or uterine contractility ultimately depends on advances of studies of the rather complex mechanisms of myometrial excitability. However, it is exactly this area of research that presents many outstanding unresolved questions. Young has clearly outlined some of such main issues hindering our further progress towards a more complete understanding of uterine contractions, including the exact roles of gap junctions, mechanisms of action potential propagation, contribution of slow waves, and functional roles of chloride and Ca^{2+}-activated K^+ channels, while expressing the hope that with enough computational power it would be possible to develop an appropriate model of human labour (Young, 2007).

Indeed, as a relevant comparison, a number of models of cardiac cells have been proposed and refined beginning from the first Noble model (Noble, 1962) allowing mechanistic insights into the nature of cardiac electrical activity. Such models include formalisation of transmembrane ionic currents, ion gradients, channel kinetics etc. culminating in a mathematical description of various electrical events. Modern concepts concerning the origin of the heartbeat include interplay between ion channel "membrane clock" mechanism and "calcium clock" mechanisms (Monfredi et al., 2013) providing the level of understanding of cardiac electrical activity and associated Ca^{2+} dynamics yet to be achieved in case of the myometrium. However, considerable progress toward this goal is being made. Here we will highlight only several of such developments that greatly contribute to our knowledge of mechanisms of myometrial excitation process and its regulation in pregnancy. At the same time it should be noted that while presently many of the experimentally recorded myometrial action potential features can be well replicated through modelling approaches, many gaps and limitations are also evident.

Thus, outlining the major advances at least briefly:

- a mathematical model of myometrial contraction has been developed to study E-C coupling, the model accounts for the operation of three Ca^{2+} control mechanisms, namely activities of voltage-gated Ca^{2+} channels, Ca^{2+} pumps, and Na^+/Ca^{2+} exchangers (Bursztyn et al., 2007);

- modelling of the uterine electrical activity in preterm labour from cellular level to surface recording was done, but reported only in an abstract from (Rihana and Marque, 2008);

- a comprehensive model based on experimentally characterised electrogenic components that can reproduce several described in the literature types of uterine action potential types (e.g., spike, plateau and short bursts of spikes) has been proposed by Tong and colleagues (Tong et al., 2011);

- its further refinement to account for the long-lasting bursting action potentials has been achieved by incorporating in the computational model delayed rectifier K^+ currents carried by potassium voltage-gated channel subfamily Q (KCNQ) and hERG channels (Tong et al., 2014), and this could explain transition from plateau-like to long-lasting bursting-type action potentials at times when uterine myocytes prepare for parturition;

- the mechanisms, which coordinate uterine contractions have been modelled based on the idea of "regional contractions" synchronised by positive feedback to produce an organ-level contraction (Young and Barendse, 2014);

- finally, the most recent advance in the modelling efforts concerns simulation of the effects of oestradiol (via reductions in Ca^{2+} and K^+ channel currents), oxytocin (via an increase in intracellular Ca^{2+} release) and the tocolytic nifedipine (via blockade of L-type Ca^{2+} channels) on action potentials and contractions, which quantitatively match the experimental data (Testrow et al., 2018).

Undoubtedly, such models provide powerful computational platforms for further experimental studies of electrogenesis and associated mechanical activity in myometrium.

Summarising, an editorial paper on quantitative analysis of uterine action potentials published by Bett (2012) concludes that "it is vital that substantial efforts are directed towards rigorous quantitative analysis and modelling of the molecular mechanisms and electrophysiological profile of the myometrium in order to move our understanding of this critical area of women's health forward" (Bett, 2012).

4.3. Regulation of Uterine Contractility

Contractility as the uterine myocytes main function undergoes complex but well-tuned control. Regulatory mechanisms of myometrial excitation and contractions comprise the myogenic, developed in the myocyte itself, paracrine that are exerted by small molecules secreted within the tissue, humoral mediated via hormones and neural fulfilled by the autonomic nervous system.

Among the myogenic mechanisms significant role belong to the focal adhesion proteins that acting together with cytoskeleton via membrane dense plaques, facilitate contractility in response to mechanical stretch (Wu et al., 2008). This pathway supposed to be involved in the onset of labour acting via ERK activation and caldesmon phosphorylation results in keeping myosin binding sites on actin filaments free thus enabling cross bridge formation and contraction development or maintaining (Li et al., 2009). Such a mechanism could be the cause of preterm labour contractions in multiple pregnancies.

Another mechanism of self-regulation developed by uterine myocytes is based on TREK-1 function. Due to outward current generation when opened and ensuing hyperpolarisation that stabilize myocytes and suppress the generation of action potentials and contractions in response to enlargement of uterine muscle fibres these channels contribute to maintaining uterine quiescence during pregnancy (Monaghan et al., 2011). Myometrial expression of TREK-1 is shown to be downregulated at the onset of labour, thus hormonal control is suggested. Buxton and co-workers have reported that post-translational modification by nitrosylation, glutathionylation or phosphorylation may result in disparate function of such channels and preterm labour. Also TREK-1 was shown to be tonically inhibited by the actin cytoskeleton and cytoskeletal reorganisation in late pregnancy could be the reason of these channels downregulation and increased contractility (Buxton et al., 2011).

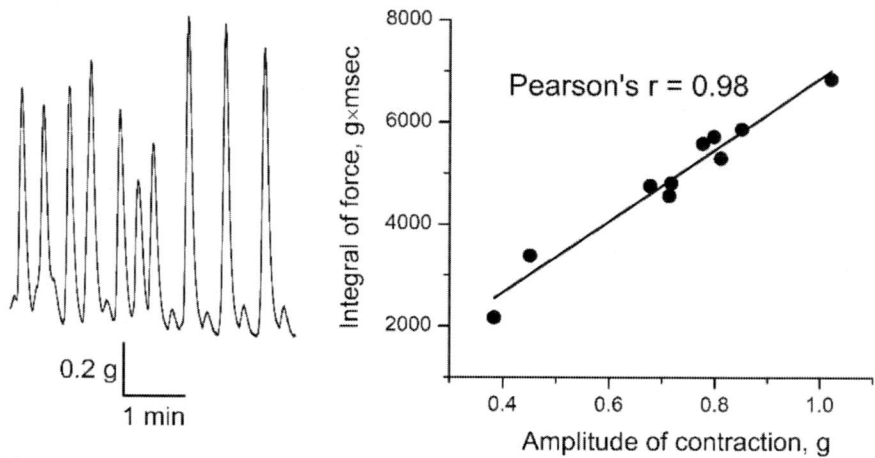

Figure 1. Typical spontaneous contractions of rat pregnant (19-23 days, n=8) myometrium (left) and linear relation between integral of force (i.e., muscle work) and amplitude of phasic contraction (right).

One more prominent way of myogenic regulation of uterine contractility is represented by mechanosensitive cation TRPV. TRPV4 channels that are expressed in human and rodent myometrium are found to participate in non-stimulated uterine contractions (Singh et al., 2015) and regulation of contractility in mouse preterm birth models (Ying et al., 2015). Our unpublished experiments on rat myometrial strips revealed decrease of the amplitude of phasic contractions by 31.7% (P<0.05) and reduction of integral of force (i.e., muscle work) by 39.3% (P<0.001) after administration of 0.3 μM of TRPV4 potent agonist GSK1016790A. (Moroz O.F., Drozd O.O., Lavryk R.T., Lazorenko I.O., Zholos A.V., unpublished). These observations are in contrast to those obtained in murine myocytes (Singh et al., 2015), but similar effect of TRPV4 stimulation was described for detrusor by Isogai and colleagues (Isogai et al., 2016). Molecular mechanism underlying diminished uterine contractility under TRPV4 stimulation could be explained by simultaneous activation of Ca^{2+}-activated potassium channels that hyperpolarize myocytes' membrane and terminate contractions. Our analysis revealed that under such experimental conditions the integral of force is linearly dependent on the contraction amplitude (Figure 1).

These channels could be activated by endogenous eicosanoids, namely epoxyeicosatrienoic acid (EET) in response to mechanical stimuli (White et al., 2016). TRPV4 sensitivity to the cell membrane stretch could be observed also in hypoosmotic environment, when cell swelling takes place. Pregnant women with hypertension disease, type II diabetes mellitus, preeclampsia and obesity usually demonstrate tissue oedema and preterm uterine contractions that could be mediated by TRPV4. Being Ca^{2+} permeable these channels, when gated, contribute to the increase of cytosolic calcium level thus facilitating force generation directly or via calcium-induced calcium release. The last remains under debates for uterine myocytes and as yet remains to be clarified (Luckas et al., 1999). Thus, TRPV4 channels due to their mechanosensitivity, calcium permeability and increased expression with the progress of gestation could be involved in molecular mechanisms of preterm labour, term labour onset and their pharmacological stimulation seems to be beneficial to enhance uterine myocytes contractility in dysfunctional labour.

Local or paracrine mechanisms of uterine myocytes regulation are under the focus of researchers because of their exclusive role mainly in the transition from quiescence to the labour contractility. In late gestation, the myometrium becomes infiltrated with immune cells (neutrophils and macrophages) that start to secrete proinflammatory cytokines. Interleukins than stimulate expression of a variety of transcription factors like tumour necrosis factor-beta (TNF-β), NF-kB, which in turn promote the expression of proteins needed for powerful labour contractility. This development of contractile phenotype includes upregulation of Cx43, OTRs expression, as well as formation of gap junctions and cytoskeleton reorganisation (Norman et al., 2007). Molecular mechanisms that provide such alteration are based on the interaction of transcription factors with promoter regions of certain genes. For instance, IL-1β induces NF-κB signalling through mitogen-activated protein kinase 1. At the same time, tumor necrosis factor alpha induces NF-κB-driven transcription of pro-labour proteins via MAPK1-independent mechanism. NF-κB activation mentioned above could be suppressed by RKIP directly. It was proved in RKIP knockdown mice where siRNA significantly increased expression and release of chemokine (C-X-C motif) ligand 1and 8 mRNA stimulated by IL-1β and TNF together with mRNA expression of PTGS2 and subsequent PGF2α release (Lappas, 2017).

Prostaglandins synthesised in pregnant uterine tissues are considered to be a key mechanism of the parturition onset. Nevertheless their synthesis in decidua and placenta are stimulated by maternal oestrogens and interleukins (Kota et al., 2013; Renthal et al., 2015). Prostaglandins, namely F2α, are found to modulate synthesis and secretion of proimflammatory cytokines by uterine SMC facilitating this way the expression of myometrial contractile proteins (Xu et al., 2015a). PGF2α together with its receptors that are Gq-protein coupled activates IP_3 and PKC and enhances myocyte contractions by increasing cytosolic Ca^{2+} level due to release of calcium from the intracellular stores and intracellular protein phosphorylation. Experiments of Xu and colleagues on cultured human pregnant myocytes revealed that PGF2α-stimulated expression of Cx43 is mediated by PKC as well as ERK (Xu et al., 2015b). Recent studies have demonstrated that impaired PG synthesis in cyclooxygenase 1 (COX-1) knockout mice leads to the delayed and dysfunctional parturition in mice. In these experiments a 24-hour delay for transcript levels of *Cx43, OTR,* and *endothelin-1* was described, but the contractility pattern of the myometrium was preserved comparatively to in term and postpartum wild-type mice. COX-1 knockout mice also had impaired cervical dilation that is supposed to occur due to both the lack of COX-1 and prolonged elevation of P_4 levels (Herington et al., 2018).

One of the intracellular pathways (for prostaglandins also) mediated by GPCR regulation of myocytes contractility could involve TRPC channels that are found to serve as receptor-operated channels for extracellular calcium entry. They are directly stimulated by phospholipase C-DAG cascade and are expressed in myometrium (Babich et al., 2004; Sanborn et al., 2005). TRPC3 channels appeared to be more highly expressed in infection-induced preterm labour model suggesting their role in cytokine mediated increase of receptor-operated calcium entry and uterine contractility (Dalrymple et al., 2004; Chen et al., 2017). In human myocytes TRPC4 channels are shown to mediate oxytocin- and prostaglandin-induced extracellular calcium influx (Ulloa et al., 2009). TRPC3 and TRPC4 channels are believed to be additionally gated by mechanical stretch and hypotonic cell swelling. Under prolonged stretch basal calcium inflow as well as the thapsigargin-provoked calcium entry were increased that was followed by the increase of TRPC3 and TRPC4 RNA content and TRPC3 protein expression (Dalrymple et al., 2007). Patch-clamp studies of rat uterine myocytes revealed that potent outward rectifying current in G protein-dependent manner was generated in response to hypotonic cell swelling and this effect was abolished by TRPC4-7 antagonist. GPCR such as those for acetylcholine or histamine are supposed to serve as mechanosensors transducing signal to TRPC4/5 channels that provide calcium entry to the cell thus enhancing uterine contractions (Chung et al., 2014).

Endocrine regulation of uterine myocytes physiology in pregnancy and parturition is provided by steroid hormones oestrogens and P4 together with peptide ones oxytocin and corticotropin-releasing hormone.

P4 is known as the main hormone of uterine quiescence and its concentration toward term becomes near maximal but due to systemic fall of P4 level that is typical for a lot of mammals or because of functional P4 withdrawal, typical for human organism, a labour occurs (Renthal et al., 2015; Nadeem et al., 2016). A mechanism that could underlie the last one is the altered ratio of different PGR (Merlino et al., 2007). It is worth mentioning here that the nuclear PRB mediate genomic effects of P4 including the inhibition of contraction-associated genes expression keeping the myometrium quiescence. In contrast, membrane associated PRA directly promote the increase of intracellular Ca^{2+} and decrease of cAMP levels that augments contractility (Mesiano, 2007). Experiments of Nadeem and colleagues revealed that during pregnancy P4 develops transcriptional repression of Cx43 in uterine myocytes by formation of homodimer complex with JUN transcription factors (AP-1 Fos/Jun family), but in labour an increased expression of Fos proteins facilitates binding of PRA to the Cx43 promoter. During parturition nuclear PRAs, being unliganded even in the presence of P4, act as transcriptional activator of Cx43. Conditions for the unliganded state of PRA are created because of intracellular metabolism of P4 by increased 20α-HSD expression in myocytes (Nadeem et al., 2016). Stimulated expression of Cx43, in turn, facilitates signal propagation between myometrial cells and the increase of uterine contractility (Nadeem et al., 2017). Some other molecular mechanisms of transition from uterine quiescence to the labour contraction are described in section "1.2. Phenotypical changes through life."

Besides direct effect on myocyte another main function of P4 is shown to be involved in the regulation of adrenoceptors expression in the myometrium. In rats pre-treated with P4 the synthesis of β_2- and α_2-adrenoceptors is increased. For α_2-adrenoceptors it is followed by coupling to G_s protein increasing cAMP level that downregulates myometrium contractility (Hajagos-Tóth et al., 2016).

Oestrogens are other hormones that play a key role in uterine myocyte physiology in pregnancy and labour. They are known to stimulate secretion of oxytocin, expression of gap junction proteins, enzymes and receptors thus preparing myometrium to forceful contractions (for review see Smith et al., 2002; Welsh et al., 2012; Ravanos et al., 2015). In humans pregnant and labouring uterus responds to high oestrogen blood level via functional changes in oestrogen receptor (ER) expression, namely ESR1 and oestrogen receptor 2 (ESR2) genes. With the labour onset the level of ESR1 rises but that of ESR2 remains unchanged (Ilicic et al., 2017). Intracellular pathways of oestrogen effect on myocytes remain unclear in many aspects but participation of ERK/ MAPK and phosphatidylinositol 3-kinase/protein kinase B (PKB/AKT) pathways and also Ca^{2+} influx and G-protein signalling were shown. Plasma membrane oestrogen receptor GPR30 activates rapid non-genomic signalling of oestradiol via its coupled G-proteins, which activate Src and the downstream ERK pathway, often by way of extracellular release of heparin-bound epidermal growth factor (HB-EGF) and transactivation of the EGF receptor (Welsh et al., 2012). These non-genomic effects being mediated by phosphorylation of MAPK and the actin-modifying HSPB 27 lead to the increased the myometrial contractile response to oxytocin (Maiti et al., 2011). Oestriol and significantly altered ratio between P4/oestriol and oestriol/oestradiol is found to be involved in rapid increase of corticotropin-releasing hormone in late pregnancy (Smith et al., 2009).

Regulation of oestrogen secretion in placenta and decidua is fulfilled by maternal corticotropin-releasing hormone and foetal adrenocorticotropin with involvement of foetal adrenal glands that secrete dehydroepiandrosteronesulphate (DHEAS), which in placenta is converted into oestrogens (Kamel, 2010).

Although the described steroid hormones are speculated to be main in pregnancy and labour, peptide hormone oxytocin remains one of the most investigated participant in the regulation of uterine myocyte contractility (Arrowsmith and Wray, 2014; Yulia and Johnson, 2014). Synthetic analogues of oxytocin are widely used to induce labour or support contractility after parturition onset and to treat postpartum haemorrhage. On the other hand, premature activation of the oxytocin system might be a cause of preterm labour. Therefore the antagonists of OTRs have been extensively examined for inhibiting preterm uterine contraction but the details of oxytocin intracellular signalling pathways are debating. Oxytocin facilitates myocyte contractility by increasing cytosolic calcium level mainly due to IP_3-mediated Ca^{2+} release from the SR that was proved in experiments with thapsigargin, where no response was detected after store depletion (Wray and Shmygol, 2007). However, such an effect of oxytocin application could trigger capacitative calcium entry via store operated calcium channels. TRPC ion channels are speculated to serve this purpose (Sanborn et al., 2005). In experiments with simultaneous recordings of luminal and cytosolic Ca^{2+} levels, only a transient fall in first one is observed at the beginning of oxytocin application, immediately followed by an increase above resting level, thus, under physiological conditions, oxytocin triggers a complex $[Ca^{2+}]_i$ response that includes initial Ca^{2+} release from the SR followed by extracellular Ca^{2+} entry. Being a G-protein coupled, OTRs are linked to phospholipase C that results in DAG formation in addition to IP_3-Ca^{2+} related pathways and alteration of calcium sensitivity of myocyte contractile apparatus.

PKC modulates MLC phosphatase activity either by direct phosphorylation or via smooth-muscle specific inhibitor CPI-17. Activation (phosphorylation) of CPI-17 inhibits the catalytic subunit of MLC phosphatase leading to increased light chain phosphorylation and enhanced contraction at a given cytosolic calcium level. In the myometrium OTR can activate alternative RhoA-GTP - ROCK pathway to phosphorylate the regulatory subunit of MLC, leading to contractions. Taking into account experimental data it could be suggested that the tonic contraction seen at the beginning of oxytocin application and altered frequency of contractions are mediated by the SR Ca^{2+}, and potentiation of contraction amplitude is achieved by sensitisation of contractile machinery to Ca^{2+} (Shmygol et al., 2006). Effects of PKC stimulation in uterine myocytes are controversial and may depend on the isoform present, the dose and duration of stimulation and may be species-specific (Arthur et al., 2007). Recently some novel roles for oxytocin and OTR have been revealed. As Kim and colleagues report, activation of oxytocin signalling stimulates cytoplasmic phospholipase A_2 (cPLA$_2$) activity and induce COX-2 expression that leads to increased PG synthesis that is found to occur first in the near cervical region thus mediating cervical ripening and dilation. These authors review the effects of oxytocin on the expression of miRNAs that in turn regulates NF-kB activity and could reduce expression of NF-kB-regulated genes including the key labour-associated genes IL-8, IL-6 and matrix metallopeptidase 9. Also, oxytocin role in myocyte physiology is represented by a receptor crosstalk. Based on the ability of OTR to form heterodimers with vasopressin, PG receptors and β-adrenergic receptors it has been suggested that binding of one oxytocin peptide to its receptor would reduce the affinity of another ligand binding to the second receptor (negative cooperative binding) (Kim et al., 2017).

There are some other endocrine factors, which significantly impact on myometrial smooth muscle cells. Among them one could mention androgens that are synthesized in corpus luteum and adrenal glands of pregnant women. Expression of AR is increased at the beginning of pregnancy, an effect that is supposed to facilitate the myocytes growth. The decrease in AR level before term directly affects insulin-like growth factor-1 receptor (IGF-1R) stability and thus down-regulates downstream cascades that IGF-1 is involved in (including phosphoinositide-3-kinase–protein kinase B/Akt (P13K/Ak)t, which is highly important in cell proliferation). Besides these androgens demonstrate relaxing effect that is rapid, suggesting non-genomic mechanism of action that underlies its possible inhibition of voltage-gated calcium channels and receptor-operated calcium influx but not the IP_3 pathway (Makieva et al., 2014).

Synthetic steroid dydrogesterone at micromolar concentrations rapidly effects on myometrial contraction, which appeared to be nongenomic, independent from PGR and mediated by voltage-dependent calcium channels (Yasuda et al., 2018).

Cholesterol, which is crucial for the formation of membrane lipid rafts and caveolae of myocytes, is shown to suppress myometrial contractility by inhibiting L-type voltage-gated calcium channels, but not by reducing calcium sensitivity. Maximal inhibitory effect is observed in myometrial strips from labouring uteri (Zhang et al., 2007).

Lactate was found to deteriorate uterine contractions both spontaneous and oxytocin-stimulated, while accumulating in amniotic fluid that could result in dysfunctional labour (dystocia). Molecular mechanisms involve the inhibition of Ca^{2+} transients because of an elevated cytosolic pH (Hanley et al., 2015). The effect diminishes with rising of pH and clinical trial has shown effectiveness of sodium bicarbonate administration on labouring outcome (Wiberg-Itzel et al., 2017).

Several substances have been described to reduce myocyte contractions and become the targets to reveal their potential pharmacological application for preterm uterine contractions. Flavonoids, among them quercetin, demonstrate inhibitory effects on phosphodiesterase and PKC that in turn increase myometrial cAMP level and promote uterine relaxation (Rezaeizadeh et al., 2016). Nitric oxide can cause S-nitrosylation of myocytes proteins that could attenuate contractions. This cGMP-independent pathway becomes of great interest as a potent pharmacological target (Ulrich et al., 2013). However, there is evidence that cGMP-dependent pathways are also involved in NO-mediated relaxation of uterine myocytes (Ergul et al., 2016) and also some reports describe the contractile effect of pharmacological substances with participation of the NO-cGMP signalling pathway (Raheja et al., 2017). Sulphides that are found recently to display various physiological effects demonstrate uterine muscle relaxation via activation of ATP-sensitive K^+ channels. Sulphides are supposed to modify the activity of other potassium channels, alter intracellular pH, phosphodiesterase activity and activity of TRP channels, as it was described for sensory nerves (Dunn et al., 2016).

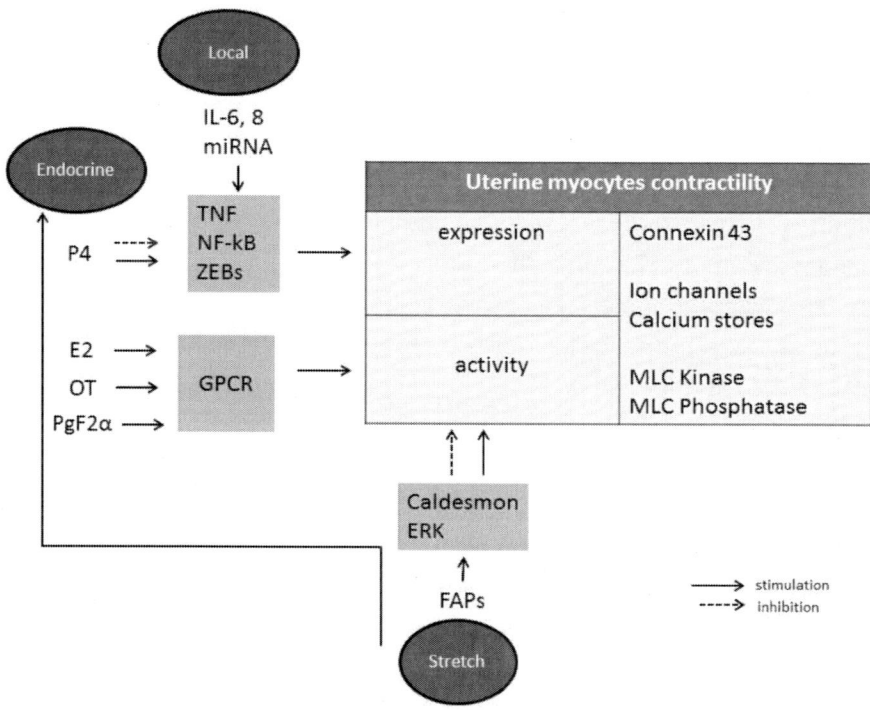

Figure 2. Schematic view on integration of regulatory mechanisms in uterine myocyte functioning (E2 – oestrogens, ERK - extracellular signal–regulated kinase, FAPs – focal adhesion proteins, GPCR - G protein-coupled receptors, IL-6, 8 – interleukins 6 and 8, NF-kB - nuclear factor –kB, OT – oxytocin, P4 – progesterone, PgF2α – prostaglandin F2α, miRNA - micro RNA, TNF – tumour necrosis factors, ZEBs - zinc finger E-box binding homeobox protein family).

Endocrine mechanisms of regulation of uterine myocyte contractility are organised into a cascade including foetal hypothalamo-pituitary-adrenal axis that affects placenta hormones secretion. This pathway appears to cooperate closely with the mechanical signalling that originates from the growing foetus (Shynlova et al., 2009). Both these mechanisms with the progress of pregnancy are directed to promote CAPs expression and labour onset (Figure 2).

Neural regulation of uterine myocytes function is found to be minimal but remains under discussion today. Being innervated with adrenergic, cholinergic and peptidergic (substance P, calcitonin gen related peptide, nitric oxide synthase-positive) fibres, uterus is shown to demonstrate their significant reduction to the end of pregnancy. Sympathetic nerves degeneration is shown for pregnant rodent and human myometrium with the progress of gestation due to oestrogens suppression of nerve growth factor (NGF) expression that is mediated via ESR1 (Latini et al., 2008). At the same time β_2 and β_3 adrenergic receptors through cAMP and cGMP signalling pathways were shown to participate in uterine quiescence or to mediate the tocolytic effects of their selective agonists (Engelhardt et al., 1997; Kaya et al., 2012). Myometrial cells of different species (mice, rats, pigs) express also muscarinic receptors type 2 and 3 (Kitazawa et al., 1999; Abdalla et al., 2004), which stimulate contractions by IP_3 dependent calcium transients and PKC-mediated calcium sensitiation (Kitazawa et al., 2008). Cholinergic innervation is affected by pregnancy to lesser extent but can be increased by oestrogens (Brauer and Smith, 2015). Neural control of uterine myocytes is considered to be involved in both keeping of pregnancy and postpartum uterine renovation.

CONCLUSION

Taking into account recent findings in the field of uterine myocyte physiology and mechanisms of maintaining uterine quiescence that are further changed to provoke contractile phenotype establishment, one can propose that uterine myocytes undergo a finely tuned regulation provided by (i) proimflammatory local factors (interleukins, various transcription factors, prostaglandins) that cause alterations in response to (ii) hormones (functional progesterone withdrawal) and (iii) mechanical stretch signals that involve focal adhesion proteins, ECM components, contractile proteins in myocytes, as well as apoptotic pathways, which acting in concert determine key phases of myometrium functioning.

Our better understanding or uterine contractility ultimately depends on advances of studies of the rather complex mechanisms of myometrial excitability and subsequent E-C coupling. Whether or not P-M coupling strictly confirms to the definition of E-C coupling is debatable, but clearly under physiological conditions these two forms (E-M and P-M) of E-C coupling co-exist and often act in synergy to produce maximally effective mechanism for force generation in smooth muscles. Mediators of these coupling modes being as potential triggers of Ca^{2+} sensitisation would be of importance in uterine (patho)physiology, but to our knowledge these have not been elucidated in the myometrium. It should be noted that while presently many of the experimentally recorded myometrial action potential features can be well replicated through modelling approaches, many gaps and limitations are also evident and require further investigations.

Uterine myocytes can transform from quiescent state during gestation to a contractile one in labour and return to resting condition in the postpartum period. Molecular mechanisms of these transitions remain to be fully clarified but several issues seem to be of prime significance: participation of TRP channels in stretch-induced and agonist-triggered myocyte responses; the role of calcium-dependent ion channels in the regulation of myocyte contractions and excitability; involvement of different elements of intracellular calcium homeostasis and transport systems like mitochondria and nucleus into regulation of contractility, and dynamic regulation of calcium sensitisation as a key point determining uterine contractility.

ACKNOWLEDGMENTS

The authors' research was supported by grants from the Department of Targeted Training of Taras Shevchenko National University of Kyiv affiliated with the National Academy of Sciences of Ukraine (grant No. 7B) and Ministry of Education and Science of Ukraine (grant No. 19BF036-01).

REFERENCES

Abdalla, F.M., Maróstica, E., Picarelli. Z.P., Abreu, L.C., Avellar, M.C.W. and Porto, C.S. (2004). Effect of estrogen on muscarinic acetylcholine receptor expression in rat myometrium. *Mol Cell Endocrinol*, 213: 139–148.

Alan, E. and Liman, N. (2018). Toll-like receptor expression patterns in the rat uterus during post partum involution. *Reprod Fertil Dev*, 30: 330-348.

Albert, A.P. and Large, W.A. (2006). Signal transduction pathways and gating mechanisms of native TRP-like cation channels in vascular myocytes. *J Physiol*, 570: 45–51.

Arango, N.A., Szotek, P.P., Manganaro, T.F., Oliva, E., Donahoe, P.K. and Teixeira, J. (2005). Conditional deletion of β-catenin in the mesenchyme of the developing mouse uterus results in a switch to adipogenesis in the myometrium. *Dev Biol*, 288: 276–283.

Arrowsmith, S., Kendrick, A., Hanley, J.-A., Noble, K. and Wray, S. (2014). Myometrial physiology - time to translate? *Exp Physiol*, 99: 495–502.

Arrowsmith, S., Kendrick, A. and Wray, S. (2010). Drugs acting on the pregnant uterus. *Obstet Gynaecol Reprod Med*, 20: 241–247.

Arrowsmith, S. and Wray, S. (2014). Oxytocin: its mechanism of action and receptor signalling in the myometrium. *J Neuroendocrinol*, 26: 356–369.

Arthur, P., Taggart, M.J. and Mitchell, B.F. (2007). Oxytocin and parturition: a role for increased myometrial calcium and calcium sensitization? *Front Biosci*, 12: 619–633.

Babich, L.G., Ku, C.-Y., Young, H.W.J., Huang, H., Blackburn, M.R. and Sanborn, B.M. (2004). Expression of capacitative calcium TrpC proteins in rat myometrium during pregnancy. *Biol Reprod*, 70: 919–924.

Bellessort, B., Bachelot, A., Heude, É., Alfama, G., Fontaine, A., Le Cardinal, M., Treier, M. and Levi, G. (2015). Role of Foxl2 in uterine maturation and function. *Hum Mol Genet*, 24: 3092–3103.

Benkusky, N.A., Fergus, D.J., Zucchero, T.M. and England, S.K. (2000). Regulation of the Ca^{2+}-sensitive domains of the maxi-K channel in the mouse myometrium during gestation. *J Biol Chem*, 275: 27712–27719.

Bett, G. (2012). Quantitative analysis of uterine action potentials. *J Genit Syst Disord*, 1: 1000e102.

Bolton, T.B., Prestwich, S.A., Zholos, A.V. and Gordienko, D.V. (1999). Excitation-contraction coupling in gastrointestinal and other smooth muscles. *Annu Rev Physiol*, 61: 85–115.

Borysova, L., Dora, K.A., Garland, C.J. and Burdyga, T. (2018). Smooth muscle gap-junctions allow propagation of intercellular Ca^{2+} waves and vasoconstriction due to Ca^{2+} based action potentials in rat mesenteric resistance arteries. *Cell Calcium*, 75: 21–29.

Boyle, A., Reddy, U.M., Landy, H.J., Huang, C.-C., Driggers, R.W. and Laughon, S.K. (2013). Primary cesarean delivery in the United States. *Obstet Gynecol*, 122: 33–40.

Brainard, A.M., Korovkina, V.P. and England, S.K. (2007). Potassium channels and uterine function. *Semin Cell Dev Biol*, 18: 332-339.

Brainard, A.M., Miller, A.J., Martens, J.R. and England, S.K. (2005). Maxi-K channels localize to caveolae in human myometrium: a role for an actin-channel-caveolin complex in the regulation of myometrial smooth muscle K^+ current. *Am J Physiol*, 289: C49–C57.

Brauer, M.M. and Burnstocks, G. (1998). Expression of β-adrenergic receptors in the rat uterus: effects of puberty and oestrogen treatment during prepubertal development. *Int J Dev Neurosci*, 16: 29–39.

Brauer, M.M. and Smith, P.G. (2015). Estrogen and female reproductive tract innervation: cellular and molecular mechanisms of autonomic neuroplasticity. *Auton Neurosci*, 187: 1–17.

Brody, J.R. and Cunha, G.R. (1989). Histologic, morphometric, and immunocytochemical analysis of myometrial development in rats and mice: I. Normal development. *Am J Anat*, 186: 1–20.

Brown, A., Cornwell, T., Korniyenko, I., Solodushko, V., Bond, C.T., Adelman, J.P. and Taylor, M.S. (2007). Myometrial expression of small conductance Ca^{2+}-activated K^+ channels depresses phasic uterine contraction. *Am J Physiol*, 292: C832-C840.

Bulbring, E., Bradimg, E., Jones, A. and Tomita, T. eds. (1981). *Smooth Muscles* (pp. 1-46). London: Edward Arnold.

Burdyga, T., Wray, S. and Noble, K. (2007). *In situ* calcium signaling: no calcium sparks detected in rat myometrium. *Ann N Y Acad Sci*, 1101: 85–96.

Bursztyn, L., Eytan, O., Jaffa, A.J. and Elad, D. (2007). Mathematical model of excitation-contraction in a uterine smooth muscle cell. *Am J Physiol*, 292: C1816–C1829.

Buxton, I.L.O., Heyman, N., Wu, Y., Barnett, S. and Ulrich, C. (2011). A role of stretch-activated potassium currents in the regulation of uterine smooth muscle contraction. *Acta Pharmacol Sin*, 32: 758–764.

Buxton, I.L.O., Singer, C.A. and Tichenor, J.N. (2010). Expression of stretch-activated two-pore potassium channels in human myometrium in pregnancy and labor. *PLoS One*, 5: e12372.

Carlsson, B. and Billig, H. (1991). Insulin-like growth factor-I gene expression during development and estrous cycle in the rat uterus. *Mol Cell Endocrinol*, 77: 175–180.

Carstem, M.E. and Miller, J.D. eds. (1990). *Uterine function: molecular and cellular aspects* (pp. 4-27). New York and London: Plenum Press.

Chang, A.N., Gao, N., Liu, Z., Huang, J., Nairn, A.C., Kamm, K.E. and Stull, J.T. (2018). The dominant protein phosphatase PP1c isoform in smooth muscle cells, PP1cβ, is essential for smooth muscle contraction. *J Biol Chem*, 293: 16677–16686.

Chawanpaiboon, S., Vogel, J.P., Moller, A.-B., Lumbiganon, P., Petzold, M., Hogan, D., Landoulsi, S., Jampathong, N., Kongwattanakul, K., Laopaiboon, M., Lewis, C., Rattanakanokchai, S., Teng, D.N., Thinkhamrop, J., Watananirun, K., Zhang, J., Zhou, W. and Gülmezoglu, A.M. (2019). Global, regional, and national estimates of levels of preterm birth in 2014: a systematic review and modelling analysis. *Lancet Glob Heal*, 7: e37–e46.

Chen, J., Zheng, D., Cui, H., Liu, S., Zhang, L. and Liu, C. (2017). Roles and mechanisms of TRPC3 and the PLCγ/PKC/CPI-17 signaling pathway in regulating parturition. *Mol Med Rep*, 17: 898–910.

Chin-Smith, E.C., Slater, D.M., Johnson, M.R. and Tribe, R.M. (2014). STIM and Orai isoform expression in pregnant human myometrium: a potential role in calcium signaling during pregnancy. *Front Physiol*, 5: 169.

Chung, S., Kim, Y.-H., Joeng, J.-H. and Ahn, D.-S. (2014). Transient receptor potential c4/5 like channel is involved in stretch-induced spontaneous uterine contraction of pregnant rat. *Korean J Physiol Pharmacol*, 18: 503–508.

Ciebiera, M., Włodarczyk, M., Zgliczyńska, M., Łukaszuk, K., Męczekalski, B., Kobierzycki, C., Łoziński, T. and Jakiel, G. (2018). The role of tumor necrosis factor α in the biology of uterine fibroids and the related symptoms. *Int J Mol Sci*, 19: E3869.

Cole, W.C., and Garfield, R.E. (1989) Ultrastructure of the Myometrium. In: Wynn, R.M., and Jollie, W.P. (eds) *Biology of the Uterus* (pp. 455-504). Boston: Springer.

Cole, W.C. and Welsh, D.G. (2011). Role of myosin light chain kinase and myosin light chain phosphatase in the resistance arterial myogenic response to intravascular pressure. *Arch Biochem Biophys*, 510: 160–173.

Colvin, C.W. and Abdullatif, H. (2013). Anatomy of female puberty: the clinical relevance of developmental changes in the reproductive system. *Clin Anat*, 26: 115–129.

Commandeur, A.E., Styer, A.K. and Teixeira, J.M. (2015). Epidemiological and genetic clues for molecular mechanisms involved in uterine leiomyoma development and growth. *Hum Reprod Update*, 21: 593–615.

Cramer, S.F., Oshri, A. and Heller, D.S. (2015). A study of myometrial growth and development. *J Pediatr Adolesc Gynecol*, 28: 387–394.

Cunha, G.R., Kurita, T., Cao, M., Shen, J., Robboy, S. and Baskin, L. (2017). Molecular mechanisms of development of the human foetal female reproductive tract. *Differentiation*, 97: 54–72.

Cunha, G.R., Robboy, S.J., Kurita, T., Isaacson, D., Shen, J., Cao, M. and Baskin, L.S. (2018). Development of the human female reproductive tract. *Differentiation*, 103: 46–65.

Curley, M., Morrison, J.J. and Smith, T.J. (2004). Analysis of Maxi-K alpha subunit splice variants in human myometrium. *Reprod Biol Endocrinol*, 2: 67.

Dalrymple, A., Mahn, K., Poston, L., Songu-Mize, E. and Tribe, R.M. (2007). Mechanical stretch regulates TRPC expression and calcium entry in human myometrial smooth muscle cells. *MHR Basic Sci Reprod Med*, 13: 171–179.

Dalrymple, A., Slater, D.M., Beech, D., Poston, L. and Tribe, R.M. (2002). Molecular identification and localization of Trp homologues, putative calcium channels, in pregnant human uterus. *Mol Hum Reprod*, 8: 946–951.

Dalrymple, A., Slater, D.M., Poston, L. and Tribe, R.M. (2004). Physiological induction of Transient Receptor Potential canonical proteins, calcium entry channels, in human myometrium: influence of pregnancy, labor, and interleukin-1β. *J Clin Endocrinol Metab*, 89: 1291–1300.

Danylovych, Y.V., Kolomiets, O. V., Danylovych, G. V.. and Kosterin, S.O. (2015). Nitric oxide as possible regulator of energy-dependent Ca^{2+} transport in mitochondria of uterine smooth muscle. *Int J Physiol Pathophysiol*, 6: 91–98.

Doheny, H.C., Lynch, C.M., Smith, T.J. and Morrison, J.J. (2005). Functional coupling of β_3 -adrenoceptors and large conductance calcium-activated potassium channels in human uterine myocytes. *J Clin Endocrinol Metab*, 90: 5786–5796.

Doring, B., Shynlova, O., Tsui, P., Eckardt, D., Janssen-Bienhold, U., Hofmann, F., Feil, S., Feil, R., Lye, S.J. and Willecke, K. (2006). Ablation of connexin43 in uterine smooth muscle cells of the mouse causes delayed parturition. *J Cell Sci*, 119: 1715–1722.

Dörr, J. and Fecher-Trost, C. (2011). TRP channels in female reproductive organs and placenta. *Adv Exp Med Biol*, 704: 909–928.

Doualla-Bell, F., Lye, S.J., Labrie, F. and Fortier, M.A. (1995). Differential expression and regulation of connexin-43 and cell-cell coupling in myocytes from the circular and longitudinal layers of bovine myometrium. *Endocrinol,* 136: 5322–5328.

Dunn, W.R., Alexander, S.P.H., Ralevic, V. and Roberts, R.E. (2016). Effects of hydrogen sulphide in smooth muscle. *Pharmacol Ther*, 158: 101–113.

Engelhardt, S., Zieger, W., Kassubek, J., Michel, M.C., Lohse, M.J. and Brodde, O.-E. (1997). Tocolytic therapy with fenoterol induces selective down-regulation of β-adrenergic receptors in human myometrium. *J Clin Endocrinol Metab*, 82: 1235–1242.

Ergul, M., Turgut, N.H., Sarac, B., Altun, A., Yildirim, Ş. and Bagcivan, I. (2016). Investigating the effects of the Rho-kinase enzyme inhibitors AS1892802 and fasudil hydrochloride on the contractions of isolated pregnant rat myometrium. *Eur J Obstet Gynecol Reprod Biol*, 202: 45–50.

Evans, W.H. and Martin, P.E.M. (2002). Gap junctions: structure and function. *Mol Membr Biol*, 19: 121–136.

Feng, L., Jayes, F., Johnson, L., Schomberg, D. and Leppert, P. (2016). Biochemical pathways and myometrial cell differentiation leading to nodule formation containing collagen and fibronectin. *Curr Protein Pept Sci*, 18: 155–166.

Ferreira, J.J., Butler, A., Stewart, R., Gonzalez-Cota, A.L., Lybaert, P., Amazu, C., Reinl, E.L., Wakle-Prabagaran, M., Salkoff, L., England, S.K. and Santi, C.M. (2019). Oxytocin can regulate myometrial smooth muscle excitability by inhibiting the Na^+-activated K^+ channel, Slo2.1. *J Physiol*, 597: 137–149.

Gao, L., Cong, B., Zhang, L. and Ni, X. (2009). Expression of the calcium-activated potassium channel in upper and lower segment human myometrium during pregnancy and parturition. *Reprod Biol Endocrinol*, 7: 27.

Gao, Y., Bayless, K.J. and Li, Q. (2014). TGFBR1 is required for mouse myometrial development. *Mol Endocrinol*, 28: 380–394.

Garfield, R.E., Merrett, D. and Grover, A.K. (1980). Gap junction formation and regulation in myometrium. *Am J Physiol*, 239: C217–C228.

Garfield, R.E., Sims, S.M., Kannan, M.S. and Daniel, E.E. (1978). Possible role of gap junctions in activation of myometrium during parturition. *Am J Physiol Physiol*, 235: C168–C179.

Gordienko, D.V. and Zholos, A.V. (2004). Regulation of muscarinic cationic current in myocytes from guinea-pig ileum by intracellular Ca^{2+} release: a central role of inositol 1,4,5-trisphosphate receptors. *Cell Calcium*, 36: 367–386.

Greenwood, I.A. and Tribe, R.M. (2014). Kv7 and Kv11 channels in myometrial regulation. *Exp Physiol*, 99: 503-509.

Gu, Y., Branham, W.S., Sheehan, D.M., Webb, P.J., Moland, C.L. and Streck, R.D. (1999). Tissue-specific expression of messenger ribonucleic acids for insulin-like growth factors and insulin-like growth factor-binding proteins during perinatal development of the rat uterus. *Biol Reprod*, 60: 1172–1182.

Guibert, C., Ducret, T. and Savineau, J.-P. (2011). Expression and physiological roles of TRP channels in smooth muscle cells. *Adv Exp Med Biol*, 704: 687–706.

Hagen, C.P., Mouritsen, A., Mieritz, M.G., Tinggaard, J., Wohlfahrt-Veje, C., Fallentin, E., Brocks, V., Sundberg, K., Jensen, L.N., Juul, A. and Main, K.M. (2015). Uterine volume and endometrial thickness in healthy girls evaluated by ultrasound (3-dimensional) and magnetic resonance imaging. *Fertil Steril*, 104: 452–459.e2.

Hajagos-Tóth, J., Bóta, J., Ducza, E., Samavati, R., Borsodi, A., Benyhe, S. and Gáspár, R. (2016). The effects of progesterone on the alpha2-adrenergic receptor subtypes in late-pregnant uterine contractions *in vitro*. *Reprod Biol Endocrinol*, 14: 33.

Hanley, J.-A., Weeks, A. and Wray, S. (2015). Physiological increases in lactate inhibit intracellular calcium transients, acidify myocytes and decrease force in term pregnant rat myometrium. *J Physiol*, 593: 4603–4614.

Herington, J.L., O'Brien, C., Robuck, M.F., Lei, W., Brown, N., Slaughter, J.C., Paria, B.C., Mahadevan-Jansen, A. and Reese, J. (2018). Prostaglandin-endoperoxide synthase 1 mediates the timing of parturition in mice despite unhindered uterine contractility. *Endocrinol*, 159: 490–505.

Hong, Sh., Kyeong, K.-S., Kim, Ch., Kim, Yc., Choi, W., Yoo, Ry., Kim, Hs., Park, Yj., Ji, Iw., Jeong, E.-H., Kim, Hs., Xu, W.-X. and Lee, Sj.

(2016). Regulation of myometrial contraction by ATP-sensitive potassium (K_{ATP}) channel via activation of SUR2B and Kir 6.2 in mouse. *J Vet Med Sci*, 78: 1153-1159.

Hu, J., Gray, C.A. and Spencer, T.E. (2004). Gene expression profiling of neonatal mouse uterine development1. *Biol Reprod*, 70: 1870–1876.

Hutchings, G., Williams, O., Cretoiu, D. and Ciontea, S.M. (2009). Myometrial interstitial cells and the coordination of myometrial contractility. *J Cell Mol Med*, 13: 4268–4282.

Ilicic, M., Zakar, T. and Paul, J.W. (2017). Modulation of progesterone receptor isoform expression in pregnant human myometrium. *Biomed Res Int*, 2017: 1–17.

Inoue, Y. and Sperelakis, N. (1991). Gestational change in Na^+ and Ca^{2+} channel current densities in rat myometrial smooth muscle cells. *Am J Physiol*, 260: C658–C663.

Isogai, A., Lee, K., Mitsui, R. and Hashitani, H. (2016). Functional coupling of TRPV4 channels and BK channels in regulating spontaneous contractions of the guinea pig urinary bladder. *Pflügers Arch*, 468: 1573–1585.

Janas, P., Kucybała, I., Radoń-Pokracka, M. and Huras, H. (2018). Telocytes in the female reproductive system: an overview of up-to-date knowledge. *Adv Clin Exp Med*, 27: 559–565.

Janjusevic, M., Greco, S., Islam, M.S., Castellucci, C., Ciavattini, A., Toti, P., Petraglia, F. and Ciarmela, P. (2016). Locostatin, a disrupter of Raf kinase inhibitor protein, inhibits extracellular matrix production, proliferation, and migration in human uterine leiomyoma and myometrial cells. *Fertil Steril*, 106: 1530–1538.e1.

Jing, C., Dongming, Z., Hong, C., Quan, N., Sishi, L. and Caixia, L. (2018). TRPC3 overexpression promotes the progression of inflammation-induced preterm labor and inhibits T cell activation. *Cell Physiol Biochem*, 45: 378–388.

Jones, K., Shmygol, A., Kupittayanant, S. and Wray, S. (2004). Electrophysiological characterization and functional importance of calcium-activated chloride channel in rat uterine myocytes. *Pflügers Arch*, 448: 36–43.

Kamel, R.M. (2010). The onset of human parturition. *Arch Gynecol Obstet*, 281: 975–982.

Kawarabayashi, T., Kishikawa, T. and Sugimori, H. (1986). Effect of oxytocin on spontaneous electrical and mechanical activities in pregnant human myometrium. *Am J Obstet Gynecol*, 155: 671–676.

Kaya, T., Karadas, B., Altun, A., Sarac, İ. and Bagcivan, I. (2012). Effects and selectivity of CL 316243, beta-3-adrenoceptor agonist, in term-pregnant rat myometrium. *Gynecol Obstet Invest*, 73: 63–69.

Kelley, C.A. and Adelstein, R.S. (1994). Characterization of isoform diversity in smooth muscle myosin heavy chains. *Can J Physiol Pharmacol*, 72: 1351–1360.

Khan, R.N., Matharoo-Ball, B., Arulkumaran, S. and Ashford, M.L.J. (2001). Potassium channels in the human myometrium. *Exp Physiol*, 86: 255–264.

Khan, R.N., Smith, S.K., Morrison, J.J. and Ashford, M.L. (1997). Ca^{2+} dependence and pharmacology of large-conductance K^+ channels in nonlabor and labor human uterine myocytes. *Am J Physiol*, 273: C1721-C1731.

Kidder, G.M. and Winterhager, E. (2015). Physiological roles of connexins in labour and lactation. *Reproduction,* 150: R129–R136.

Kilarski, W.M., Rothery, S., Roomans, G.M., Ulmsten, U., Rezapour, M., Stevenson, S., Coppen, S.R., Dupont, E. and Severs, N.J. (2001). Multiple connexins localized to individual gap-junctional plaques in human myometrial smooth muscle. *Microsc Res Tech*, 54: 114–122.

Kim, S.H., Bennett, P.R. and Terzidou, V. (2017). Advances in the role of oxytocin receptors in human parturition. *Mol Cell Endocrinol*, 449: 56–63.

Kitazawa, T., Hirama, R., Masunaga, K., Nakamura, T., Asakawa, K., Cao, J., Teraoka, H., Unno, T., Komori, S., Yamada, M., Wess, J. and Taneike, T. (2008). Muscarinic receptor subtypes involved in carbachol-induced contraction of mouse uterine smooth muscle. *Naunyn Schmiedebergs Arch Pharmacol*, 377: 503–513.

Kitazawa, T., Uchiyama, F., Hirose, K. and Taneike, T. (1999). Characterization of the muscarinic receptor subtype that mediates the contractile response of acetylcholine in the swine myometrium. *Eur J Pharmacol*, 367: 325–334.

Kizub, I. V., Pavlova, O.O., Johnson, C.D., Soloviev, A.I. and Zholos, A. V. (2010). Rho kinase and protein kinase C involvement in vascular smooth muscle myofilament calcium sensitization in arteries from diabetic rats. *Br J Pharmacol*, 159: 1724–1731.

Klein, U., Gimpl, G, and Fahrenholz, F. (1995). Alteration of the myometrial plasma membrane cholesterol content with beta-cyclodextrin modulates the binding affinity of the oxytocin receptor. *Biochemistry*, 34: 13784–13793.

Knock, G.A. and Aaronson, P.I. (1999). Calcium antagonistic properties of the cyclooxygenase-2 inhibitor nimesulide in human myometrial myocytes. *Br J Pharmacol*, 127: 1470–1478.

Kolomiiets, O.V., Danylovych, I.V. and Danylovych, H.V. (2014a). [H$^+$-Ca^{2+}-exchanger in the myometrium mitochondria: modulation of exogenous and endogenous compounds]. *Fiziol Zh*, 60: 33–42.

Kolomiiets, O.V., Danylovych, I.V., Danylovych, H.V. and Kosterin, S.O. (2014b). [Ca^{2+}/H$^+$-exchange in myometrium mitochondria]. *Ukr Biochem J*, 86: 41–48.

Kopylova, G., Nabiev, S., Nikitina, L., Shchepkin, D. and Bershitsky, S. (2016). The properties of the actin-myosin interaction in the heart muscle depend on the isoforms of myosin but not of α-actin. *Biochem Biophys Res Commun*, 476: 648–653.

Kosterin, S.A., Bratkova, N.F. and Kurskiĭ, M.D. (1985). [The role of sarcolemma and mitochondria in calcium-dependent control of myometrium relaxation]. *Biokhimiia*, 50: 1350–1361.

Kota, S., Gayatri, K., Jammula, S., Kota, S., Krishna, S.V.S., Meher, L. and Modi, K. (2013). Endocrinology of parturition. *Indian J Endocrinol Metab*, 17: 50-59.

Krebsbach, P.H. and Villa-Diaz, L.G. (2017). The role of integrin α6 (CD49f) in stem cells: more than a conserved biomarker. *Stem Cells Dev*, 26: 1090–1099.

Kuo, I.Y. and Ehrlich, B.E. (2015). Signalling in muscle contraction. *Cold Spring Harb Perspect Biol*, 7: a006023.

Labyntseva, R.D., Bevza, O.V., Lytvyn, K.V., Borovyk, M.O., Rodik, R.V., Kalchenko, V.I. and Kosterin, S.O. (2016). Calix[4]arene C-90 and its analogs activate ATPase of the myometrium myosin subfragment-1. *Ukr Biochem J*, 88: 48–61.

Lappas, M. (2017). RKIP is decreased in laboring myometrium and modulates inflammation-induced pro-labor mediators. *Reproduction*, 153: 545–553.

Lartey, J., Taggart, J., Robson, S. and Taggart, M. (2016). Altered expression of human smooth muscle myosin phosphatase targeting (MYPT) isovariants with pregnancy and labor. *PLoS One*, 11: e0164352.

Latini, C., Frontini, A., Morroni, M., Marzioni, D., Castellucci, M. and Smith, P.G. (2008). Remodeling of uterine innervation. *Cell Tissue Res*, 334: 1–6.

Lee, S.-E., Ahn, D.-S. and Lee, Y.-H. (2009). Role of T-type Ca channels in the spontaneous phasic contraction of pregnant rat uterine smooth muscle. *Korean J Physiol Pharmacol*, 13: 241–249.

Lee, Y., Sooranna, S.R., Terzidou, V., Christian, M., Brosens, J., Huhtinen, K., Poutanen, M., Barton, G., Johnson, M.R. and Bennett, P.R. (2012). Interactions between inflammatory signals and the progesterone receptor in regulating gene expression in pregnant human uterine myocytes. *J Cell Mol Med*, 16: 2487–2503.

Li, Q., Agno, J.E., Edson, M.A., Nagaraja, A.K., Nagashima, T. and Matzuk, M.M. (2011). Transforming growth factor β receptor type 1 is essential for female reproductive tract integrity and function. *PLoS Genet*, 7: e1002320.

Li, Y., Gallant, C., Malek, S. and Morgan, K.G. (2007). Focal adhesion signaling is required for myometrial ERK activation and contractile phenotype switch before labor. *J Cell Biochem*, 100: 129–140.

Li, Y., Reznichenko, M., Tribe, R.M., Hess, P.E., Taggart, M., Kim, H., DeGnore, J.P., Gangopadhyay, S. and Morgan, K.G. (2009). Stretch activates human myometrium via ERK, caldesmon and focal adhesion signaling. *PLoS One*, 4: e7489.

Lorca, R.A., Prabagaran, M. and England, S.K. (2014). Functional insights into modulation of BK_{Ca} channel activity to alter myometrial contractility. *Front Physiol*, 5: 289.

Luckas, M.J., Taggart, M.J. and Wray, S. (1999). Intracellular calcium stores and agonist-induced contractions in isolated human myometrium. *Am J Obstet Gynecol*, 181: 468–476.

MacKenzie, L.W. and Garfield, R.E. (1985). Hormonal control of gap junctions in the myometrium. *Am J Physiol*, 248: C296–C308.

MacPhee, D.J. and Miskiewicz, E.I. (2017). The potential functions of small heat shock proteins in the uterine musculature during pregnancy. *Adv Anat Embryol Cell Biol,* 222: 95-116.

Maiti, K., Paul, J.W., Read, M., Chan, E.C., Riley, S.C., Nahar, P. and Smith, R. (2011). G-1-activated membrane estrogen receptors mediate increased contractility of the human myometrium. *Endocrinol*, 152: 2448–2455.

Makieva, S., Saunders, P.T.K. and Norman, J.E. (2014). Androgens in pregnancy: roles in parturition. *Hum Reprod Update*, 20: 542–559.

Marchi, S. and Pinton, P. (2014). The mitochondrial calcium uniporter complex: molecular components, structure and physiopathological implications. *J Physiol*, 592: 829–839.

Martin, C., Arnaudeau, S., Jmari, K., Rakotoarisoa, L., Sayet, I., Dacquet, C., Mironneau, C. and Mironneau, J. (1990). Identification and properties of voltage-sensitive sodium channels in smooth muscle cells from pregnant rat myometrium. *Mol Pharmacol*, 38: 667–673.

Maruyama, T., Masuda, H., Ono, M., Kajitani, T. and Yoshimura, Y. (2010). Human uterine stem/progenitor cells: their possible role in uterine physiology and pathology. *Reproduction*, 140: 11–22.

Mas, A., Cervelló, I., Fernández-Álvarez, A., Faus, A., Díaz, A., Burgués, O., Casado, M. and Simón, C. (2015). Overexpression of the truncated form of High Mobility Group A proteins (HMGA2) in human myometrial cells induces leiomyoma-like tissue formation. *MHR Basic Sci Reprod Med*, 21: 330–338.

Matharoo-Ball, B., Ashford, M.L.J., Arulkumaran, S. and Khan, R.N. (2003). Down-regulation of the α- and β-subunits of the calcium-activated potassium channel in human myometrium with parturition. *Biol Reprod*, 68: 2135–2141.

Mehasseb, M.K., Bell, S.C. and Habiba, M.A. (2009). The effects of tamoxifen and estradiol on myometrial differentiation and organization during early uterine development in the CD1 mouse. *Reproduction*, 138: 341–350.

Merlino, A.A., Welsh, T.N., Tan, H., Yi, L.J., Cannon, V., Mercer, B.M. and Mesiano, S. (2007). Nuclear progesterone receptors in the human pregnancy myometrium: evidence that parturition involves functional progesterone withdrawal mediated by increased expression of progesterone receptor-A. *J Clin Endocrinol Metab*, 92: 1927–1933.

Mesiano, S. (2004). Myometrial progesterone responsiveness and the control of human parturition. *J Soc Gynecol Investig*, 11: 193–202.

Mesiano S, (2007). Myometrial progesterone responsiveness. *Semin Reprod Med*, 25: 005–013.

Miftahof, R. and Nam, H.G. (2011). *Biomechanics of the Gravid Human Uterus* (pp. 1-17). Heidelberg Dordrecht London New York: Springer.

Modzelewska, B., Kostrzewska, A., Sipowicz, M., Kleszczewski, T. and Batra, S. (2003). Apamin inhibits NO-induced relaxation of the spontaneous contractile activity of the myometrium from non-pregnant women. *Reprod Biol Endocrinol*, 1: 8.

Monaghan, K., Baker, S.A., Dwyer, L., Hatton, W.C., Sik Park, K., Sanders, K.M. and Koh, S.D. (2011). The stretch-dependent potassium channel TREK-1 and its function in murine myometrium. *J Physiol*, 589: 1221–1233.

Monfredi, O., Maltsev, V.A. and Lakatta, E.G. (2013). Modern concepts concerning the origin of the heartbeat. *Physiol*, 28, 74–92.

Morgan, J.M., De Smedt, H. and Gillespie, J.I. (1996). Identification of three isoforms of the InsP$_3$ receptor in human myometrial smooth muscle. *Pflügers Arch*, 431: 697–705.

Morgan, K.G. (2014). The importance of the smooth muscle cytoskeleton to preterm labour. *Exp Physiol*, 99: 525–529.

Murtazina, D.A., Chung, D., Ulloa, A., Bryan, E., Galan, H.L. and Sanborn, B.M. (2011). TRPC1, STIM1, and ORAI influence signal-regulated intracellular and endoplasmic reticulum calcium dynamics in human myometrial cells. *Biol Reprod*, 85: 315–326.

Nadeem, L., Shynlova, O., Matysiak-Zablocki, E., Mesiano, S., Dong, X. and Lye, S. (2016). Molecular evidence of functional progesterone withdrawal in human myometrium. *Nat Commun*, 7: 11565.

Nadeem, L., Shynlova, O., Mesiano, S. and Lye, S. (2017). Progesterone via its type-A receptor promotes myometrial gap junction coupling. *Sci Rep*, 7: 13357.

Nelson, M.T., Cheng, H., Rubart, M., Santana, L.F., Bonev, A.D., Knot, H.J. and Lederer, W.J. (1995). Relaxation of arterial smooth muscle by calcium sparks. *Science*, 270: 633–637.

Noble, D. (1962). A modification of the Hodgkin-Huxley equations applicable to Purkinje fibre action and pacemaker potentials. *J Physiol*, 160: 317–352.

Noble, D., Borysova, L., Wray, S. and Burdyga, T. (2014). Store-operated Ca^{2+} entry and depolarization explain the anomalous behaviour of myometrial SR: effects of SERCA inhibition on electrical activity, Ca^{2+} and force. *Cell Calcium*, 56: 188–194.

Noble, K., Zhang, J. and Wray, S. (2006). Lipid rafts, the sarcoplasmic reticulum and uterine calcium signalling: an integrated approach. *J Physiol*, 570: 29–35.

Norman, J.E., Bollapragada, S., Yuan, M. and Nelson, S.M. (2007). Inflammatory pathways in the mechanism of parturition. *BMC Pregnancy Childbirth*, 7: S7.

Nothnick, W.B. (2016). Non-coding RNAs in uterine development, function and disease. *Adv Exp Med Biol*, 886: 171-189.

Ohkubo, T., Kawarabayashi, T., Inoue, Y. and Kitamura, K. (2005). Differential expression of L- and T-type calcium channels between longitudinal and circular muscles of the rat myometrium during pregnancy. *Gynecol Obstet Invest*, 59: 80–85.

Ohya, Y. and Sperelakis, N. (1989). Fast Na^+ and slow Ca^{2+} channels in single uterine muscle cells from pregnant rats. *Am J Physiol*, 257: C408–C412.

Ono, M., Maruyama, T., Masuda, H., Kajitani, T., Nagashima, T., Arase, T., Ito, M., Ohta, K., Uchida, H., Asada, H., Yoshimura, Y., Okano, H. and Matsuzaki, Y. (2007). Side population in human uterine myometrium displays phenotypic and functional characteristics of myometrial stem cells. *Proc Natl Acad Sci*, 104: 18700–18705.

Parkington, H.C. and Coleman, H.A. (1990). The role of membrane potential in the control of uterine motility. In *Uterine Function* (pp. 195–248). Boston: Springer US.

Parkington, H.C., Stevenson, J., Tonta, M.A., Paul, J., Butler, T., Maiti, K., Chan, E.-C., Sheehan, P.M., Brennecke, S.P., Coleman, H.A. and Smith, R. (2014). Diminished hERG K$^+$ channel activity facilitates strong human labour contractions but is dysregulated in obese women. *Nat Commun*, 5: 4108.

Parkington, H.C., Tonta, M.A., Brennecke, S.P. and Coleman, H.A. (1999). Contractile activity, membrane potential, and cytoplasmic calcium in human uterine smooth muscle in the third trimester of pregnancy and during labor. *Am J Obstet Gynecol*, 181: 1445–1451.

Prunskaite-Hyyryläinen, R., Skovorodkin, I., Xu, Q., Miinalainen, I., Shan, J. and Vainio, S.J. (2016). Wnt4 coordinates directional cell migration and extension of the Müllerian duct essential for ontogenesis of the female reproductive tract. *Hum Mol Genet*, 25: 1059–1073.

Raheja, R., Gupta, H., Pandey, U. and Deshpande, S.B. (2017). Lignocaine augments the in-vitro uterine contractions involving NO-guanylyl cyclase dependent mechanisms. *Life Sci*, 190: 52–57.

Ravanos, K., Dagklis, T., Petousis, S., Margioula-Siarkou, C., Prapas, Y. and Prapas, N. (2015). Factors implicated in the initiation of human parturition in term and preterm labor: a review. *Gynecol Endocrinol*, 31: 679–683.

Renthal, N.E., Williams, K.C., Montalbano, A.P., Chen, C.-C., Gao, L. and Mendelson, C.R. (2015). Molecular regulation of parturition: a myometrial perspective. *Cold Spring Harb Perspect Med*, 5: a023069.

Rezaeizadeh, G., Hantoushzadeh, S., Ghiasi, S., Nikfar, S. and Abdollahi, M. (2016). A systematic review of the uterine relaxant effect of herbal sources. *Curr Pharm Biotechnol*, 17: 934–948.

Rihana, S. and Marque, C. (2008). Preterm labor - modeling the uterine electrical activity from cellular level to surface recording. In: *30th Annual International Conference of the IEEE Engineering in Medicine and Biology Society* (pp. 3726–3729). IEEE.

Robboy, S.J., Kurita, T., Baskin, L. and Cunha, G.R. (2017). New insights into human female reproductive tract development. *Differentiation*, 97: 9–22.

Robinson, R.S., Mann, G.E., Gadd, T.S., Lamming, G.E. and Wathes, D.C. (2000). The expression of the IGF system in the bovine uterus throughout the oestrous cycle and early pregnancy. *J Endocrinol*, 165: 231–243.

Rodriguez, A., Tripurani, S.K., Burton, J.C., Clementi, C., Larina, I. and Pangas, S.A. (2016). SMAD signalling is required for structural integrity of the female reproductive tract and uterine function during early pregnancy in mice. *Biol Reprod*, 95: 44–44.

Rovner, A.S., Fagnant, P.M., Lowey, S. and Trybus, K.M. (2002). The carboxyl-terminal isoforms of smooth muscle myosin heavy chain determine thick filament assembly properties. *J Cell Biol*, 156: 113–124.

Sanborn, B.M. (2000). Relationship of ion channel activity to control of myometrial calcium. *J Soc Gynecol Investig*, 7: 4–11.

Sanborn, B.M., Ku, C.-Y., Shlykov, S. and Babich, L. (2005). Molecular signaling through G-protein-coupled receptors and the control of intracellular calcium in myometrium. *J Soc Gynecol Investig*, 12: 479–487.

Seda, M., Pinto, F.M., Wray, S., Cintado, C.G., Noheda, P., Buschmann, H. and Candenas, L. (2007). Functional and molecular characterization of voltage-gated sodium channels in uteri from nonpregnant rats. *Biol Reprod*, 77: 855–863.

Shmygol, A., Gullam, J., Blanks, A. and Thornton, S. (2006). Multiple mechanisms involved in oxytocin-induced modulation of myometrial contractility. *Acta Pharmacol Sin*, 27: 827–832.

Shynlova, O., Tsui, P., Dorogin, A., Langille, B.L. and Lye, S.J. (2007). Insulin-like growth factors and their binding proteins define specific phases of myometrial differentiation during pregnancy in the rat. *Biol Reprod*, 76: 571–578.

Shynlova, O., Tsui, P., Jaffer, S. and Lye, S.J. (2009). Integration of endocrine and mechanical signals in the regulation of myometrial functions during pregnancy and labour. *Eur J Obstet Gynecol Reprod Biol*, 144: S2–S10.

Singh, V., Ram, M., Kandasamy, K., Thangamalai, R., Choudhary, S., Dash, J.R., Kumar, D., Parida, S., Singh, T.U. and Mishra, S.K. (2015). Molecular and functional characterization of TRPV4 channels in pregnant and nonpregnant mouse uterus. *Life Sci*, 122: 51–58.

Smirnov, S.V., Zholos, A.V. and Shuba, M.F. (1992). Potential-dependent inward currents in single isolated smooth muscle cells of the rat ileum. *J Physiol*, 454: 549-571.

Smith, R., Mesiano, S. and McGrath, S. (2002). Hormone trajectories leading to human birth. *Regul Pept*, 108: 159–164.

Smith, R., Smith, J.I., Shen, X., Engel, P.J., Bowman, M.E., McGrath, S.A., Bisits, A.M., McElduff, P., Giles, W.B. and Smith, D.W. (2009). Patterns of plasma corticotropin-releasing hormone, progesterone, estradiol, and estriol change and the onset of human labor. *J Clin Endocrinol Metab*, 94: 2066–2074.

Smith, R.C., McClure, M.C., Smith, M.A., Abel, P.W. and Bradley, M.E. (2007). The role of voltage-gated potassium channels in the regulation of mouse uterine contractility. *Reprod Biol Endocrinol*, 5: 41.

Somara, S., Pang, H. and Bitar, K.N. (2005). Agonist-induced association of tropomyosin with protein kinase Cα in colonic smooth muscle. *Am J Physiol*, 288: G268–G276.

Spencer, T.E., Hayashi, K., Hu, J. and Carpenter, K.D. (2005). Comparative developmental biology of the mammalian uterus. In Gerald P. Schatten (Ed.) *Current topics in developmental biology* (pp. 85–122). Academic Press. Stewart, E.A. (2001). Uterine fibroids. *Lancet*, 357: 293–298.

Taggart, M.J., Blanks, A., Kharche, S., Holden, A., Wang, B. and Zhang, H. (2007). Towards understanding the myometrial physiome: approaches for the construction of a virtual physiological uterus. *BMC Pregnancy Childbirth*, 7: S3.

Taggart, M.J. and Morgan, K.G. (2007). Regulation of the uterine contractile apparatus and cytoskeleton. *Semin Cell Dev Biol*, 18: 296–304.

Testrow, C.P., Holden, A.V., Shmygol, A. and Zhang, H. (2018). A computational model of excitation and contraction in uterine myocytes from the pregnant rat. *Sci Rep*, 8: 9159.

Tong, W.C., Choi, C.Y., Karche, S., Holden, A.V., Zhang, H. and Taggart, M.J. (2011). A computational model of the ionic currents, Ca^{2+} dynamics and action potentials underlying contraction of isolated uterine smooth muscle. *PLoS One*, 6: e18685.

Tong, W.C., Tribe, R.M., Smith, R. and Taggart, M.J. (2014). Computational modeling reveals key contributions of KCNQ and hERG currents to the malleability of uterine action potentials underpinning labor. *PLoS One*, 9: e114034.

Tribe, R.M. (2001). Regulation of human myometrial contractility during pregnancy and labour: are calcium homeostatic pathways important? *Exp Physiol*, 86: 247-254.

Tribe, R.M., Moriarty, P. and Poston, L. (2000). Calcium homeostatic pathways change with gestation in human myometrium. *Biol Reprod*, 63: 748–755.

Turi, A., Kiss, A.L. and Müllner, N. (2001). Estrogen downregulates teh number of caveolae and the level of caveolin in uterine smooth muscle. *Cell Biol Int*, 25: 785–794.

Ulloa, A., Gonzales, A.L., Zhong, M., Kim, Y.-S., Cantlon, J., Clay, C., Ku, C.-Y., Earley, S. and Sanborn, B.M. (2009). Reduction in TRPC4 expression specifically attenuates G-protein coupled receptor-stimulated increases in intracellular calcium in human myometrial cells. *Cell Calcium*, 46: 73–84.

Ulrich, C., Quilici, D.R., Schlauch, K.A. and Buxton, I.L.O. (2013). The human uterine smooth muscle *S*-nitrosoproteome fingerprint in pregnancy, labor, and preterm labor. *Am J Physiol*, 305: C803–C816.

Varga, I., Klein. M,, Urban, L., Danihel, L., Polak, S. and Danihel, L. (2018). Recently discovered interstitial cells "telocytes" as players in the pathogenesis of uterine leiomyomas. *Med Hypotheses*, 110: 64–67.

Venkatachalam, K. and Montell, C. (2007). TRP channels. *Annu Rev Biochem*, 76: 387–417.

Vishnyakova, P.A., Tarasova, N.V., Volodina, M.A., Tsvirkun, D.V., Sukhanova, I.A., Kurchakova, T.A., Kan, N.E., Medzidova, M.K., Sukhikh, G.T. and Vysokikh. M.Y. (2019). Gestation age-associated dynamics of mitochondrial calcium uniporter subunits expression in feto-maternal complex at term and preterm delivery. *Sci Rep*, 9: 5501.

Wakle-Prabagaran, M., Lorca, R.A., Ma, X., Stamnes, S.J., Amazu, C., Hsiao, J.J., Karch, C.M., Hyrc, K.L., Wright, M.E. and England, S.K. (2016). BK_{Ca} channel regulates calcium oscillations induced by alpha-2-macroglobulin in human myometrial smooth muscle cells. *Proc Natl Acad Sci USA*, 113: E2335–E2344.

Wang, S.Y., Yoshino, M., Sui, J.L., Wakui, M., Kao, P.N. and Kao, C.Y. (1998). Potassium currents in freshly dissociated uterine myocytes from nonpregnant and late-pregnant rats. *J Gen Physiol*, 112: 737–756.

Wang, Y., Jia, Y., Franken, P., Smits, R., Ewing, P.C., Lydon, J.P., DeMayo, F.J., Burger, C.W., Anton Grootegoed, J., Fodde, R. and Blok, L.J. (2011). Loss of APC function in mesenchymal cells surrounding the Müllerian duct leads to myometrial defects in adult mice. *Mol Cell Endocrinol*, 341: 48–54.

Wellman, G.C., Nathan, D.J., Saundry, C.M., Perez, G., Bonev, A.D., Penar, P.L., Tranmer, B.I. and Nelson, M.T. (2002). Ca^{2+} sparks and their function in human cerebral arteries. *Stroke*, 33: 802–808.

Welsh, T., Johnson, M., Yi, L., Tan, H., Rahman, R., Merlino, A., Zakar, T. and Mesiano, S. (2012). Estrogen receptor (ER) expression and function in the pregnant human myometrium: estradiol via ERα activates ERK1/2 signaling in term myometrium. *J Endocrinol*, 212: 227–238.

White, J.P.M., Cibelli, M., Urban, L., Nilius, B., McGeown, J.G. and Nagy, I. (2016). TRPV4: molecular conductor of a diverse orchestra. *Physiol Rev*, 96: 911–973.

Wiberg-Itzel, E., Wray, S. and Åkerud, H. (2017). A randomized controlled trial of a new treatment for labor dystocia. *J Matern Foetal Neonatal Med*, 31: 2237-2244.

Williams, K.C., Renthal, N.E., Condon, J.C., Gerard, R.D. and Mendelson, C.R. (2012). MicroRNA-200a serves a key role in the decline of progesterone receptor function leading to term and preterm labor. *Proc Natl Acad Sci USA*, 109: 7529–7534.

Wray, S. (2007). Insights into the uterus. *Exp Physiol*, 92: 621–631.

Wray, S. and Shmygol, A. (2007). Role of the calcium store in uterine contractility. *Semin Cell Dev Biol*, 18: 315–320.

Wu, X., Morgan, K.G., Jones, C.J., Tribe, R.M. and Taggart, M.J. (2008). Myometrial mechanoadaptation during pregnancy: implications for smooth muscle plasticity and remodelling. *J Cell Mol Med*, 12: 1360–1373.

Wynn, R.M. and Jollie, W.P. (Eds). (1989). *Biology of the Uterus*, (pp.355-503). New York and London: Plemium Medical Book Company.

Xu, C., Liu, W., You, X., Leimert, K., Popowycz, K., Fang, X., Wood, S.L., Slater, D.M., Sun, Q., Gu, H., Olson, D.M. and Ni, X. (2015*a*). PGF$_{2\alpha}$ modulates the output of chemokines and pro-inflammatory cytokines in myometrial cells from term pregnant women through divergent signaling pathways. *Mol Hum Reprod*, 21: 603–614.

Xu, C., You, X., Liu, W., Sun, Q., Ding, X., Huang, Y. and Ni, X. (2015*b*). Prostaglandin F2α regulates the expression of uterine activation proteins via multiple signalling pathways. *Reproduction*, 149: 139–146.

Yang, M., Gupta, A., Shlykov, S.G., Corrigan, R., Tsujimoto, S. and Sanborn, B.M. (2002). Multiple Trp isoforms implicated in capacitative calcium entry are expressed in human pregnant myometrium and myometrial cells. *Biol Reprod*, 67: 988–994.

Yasuda, K., Sumi, G., Murata, H., Kida, N., Kido, T. and Okada, H. (2018). The steroid hormone dydrogesterone inhibits myometrial contraction independently of the progesterone/progesterone receptor pathway. *Life Sci*, 207: 508–515.

Ying, L., Becard, M., Lyell, D., Han, X., Shortliffe, L., Husted, C.I., Alvira, C.M. and Cornfield, D.N. (2015). The transient receptor potential vanilloid 4 channel modulates uterine tone during pregnancy. *Sci Transl Med*, 7: 319ra204.

Yoshino, M., Wang, S.Y. and Kao, C.Y. (1997). Sodium and calcium inward currents in freshly dissociated smooth myocytes of rat uterus. *J Gen Physiol*, 110: 565–577.

Young, R.C. (2007). Myocytes, myometrium, and uterine contractions. *Ann N Y Acad Sci*, 1101: 72–84.

Young, R.C. and Barendse, P. (2014). Linking myometrial physiology to intrauterine pressure; how tissue-level contractions create uterine contractions of labor. *PLoS Comput Biol*, 10: e1003850.

Young, R.C. and Herndon-Smith, L. (1991). Characterization of sodium channels in cultured human uterine smooth muscle cells. *Am J Obstet Gynecol*, 164: 175–181.

Young, R.C., Smith, L.H. and McLaren, M.D. (1993). T-type and L-type calcium currents in freshly dispersed human uterine smooth muscle cells. *Am J Obstet Gynecol*, 169: 785–792.

Young, R.C. and Zhang, P. (2004). Functional separation of deep cytoplasmic calcium from subplasmalemmal space calcium in cultured human uterine smooth muscle cells. *Cell Calcium*, 36: 11–17.

Yue, Z., Xie, J., Yu, A.S., Stock, J., Du, J. and Yue, L. (2015). Role of TRP channels in the cardiovascular system. *Am J Physiol*, 308: H157-82.

Yulia, A. and Johnson, M.R. (2014). Myometrial oxytocin receptor expression and intracellular pathways. *Minerva Ginecol*, 66: 267–280.

Zhang, J,, Kendrick, A., Quenby, S. and Wray, S. (2007). Contractility and calcium signaling of human myometrium are profoundly affected by cholesterol manipulation: implications for labor? *Reprod Sci*, 14: 456–466.

Zheng, D., Zhang, L., Na, Q., Liu, S., Zhuang, Y., Lv, Y. and Liu, C. (2016). Enhanced expression of transient receptor potential channel 3 in uterine smooth muscle tissues of lipopolysaccharide-induced preterm delivery mice. *Iran J Basic Med Sci*, 19: 567–572.

Zholos, A.V., Baidan, L.V. and Shuba, M.F. (1991). Properties of the late transient outward current in isolated intestinal smooth muscle cells of the guinea-pig. *J Physiol*, 443: 555-574.

Zholos, A.V., Baidan, L.V. and Shuba, M.F. (1992). Some properties of Ca^{2+}-induced Ca^{2+} release mechanism in single visceral smooth muscle cell of the guinea-pig. *J Physiol*, 457: 1-25.

Zhou, X.-B., Wang, G.-X., Ruth, P., Hüneke, B. and Korth, M. (2000). BK_{Ca} channel activation by membrane-associated cGMP kinase may contribute to uterine quiescence in pregnancy. *Am J Physiol*, 279: C1751–C1759.

In: Advances in Medicine and Biology   ISBN: 978-1-53616-181-6
Editor: Leon V. Berhardt   © 2019 Nova Science Publishers, Inc.

Chapter 3

TARGETED THERAPIES IN BREAST CANCER

Sara Charmsaz, Stephen Keelan, Fiona Bane,
Michael Flanagan, Arnold K. Hill and Leonie S. Young[*]
Endocrine Oncology Research Group, Department of Surgery,
Royal College of Surgeons in Ireland, Dublin, Dublin, Ireland

ABSTRACT

Recent advances in targeted therapy have decreased the morbidity and enhanced the quality of life of cancer patients. The use of peptides and monoclonal antibodies as targeted therapies has received considerable attention in recent years, establishing this mode of treatment as an important therapeutic strategy. Breast cancer is the most common type of cancer in women and one of the few malignancies in which tumour heterogeneity have been successfully used for molecular classification and therapeutic intervention. Advances in molecular biology including genomics, epi-genomics and transcriptomics have resulted in the identification of better treatments that have been tailored to target specific pathophysiology. Here we provide an overview of new developments in breast cancer targeted therapies and discuss future prospects for directed therapeutic strategies.

[*] Corresponding Author's E-mail: lyoung@rcsi.com.

CANCER AND TARGETED THERAPIES

In 2000, Hanahan and Weinberg introduced the hallmark of cancer where they described six crucial alterations that result in cancer cell growth. These included self-sufficiency in growth signals, insensitivity to anti-growth signals, evading apoptosis, limitless replicative potential, sustained angiogenesis, and tissue invasion and metastasis (Hanahan and Weinberg 2000). Eleven years later they completed the hallmark with the addition of another two alterations: firstly, deregulation in cellular metabolism and avoidance of immune destruction, and secondly enabling characteristics including genomic instability, mutation, and tumour-promoting inflammation (Hanahan and Weinberg 2011). Definition of the cancer hallmarks together with new biological advancements in recent years has resulted in the development of novel therapeutics that interfere with at least one of these acquired capabilities for cancer cell growth and progression and have led to a new era of targeted cancer treatments. Identification of these therapeutic targets, unlike the classic chemotherapeutics, is dependent on an in-depth understanding of the molecular changes that are specific to cancer cells. Molecular targeted therapy focuses not only on changes that lead to malignant progression, but also takes into consideration the tumour's unique clinical and molecular features. Recent advances in genomics have improved our understanding of tumour heterogeneity and lead to individualised treatment strategies.

Targeted therapeutics have been implemented in many different malignancies including glioma, colorectal, lung, pancreatic, ovarian as well as lymphoma, leukaemia, multiple myeloma and breast cancer. Unlike classic chemotherapy and radiotherapy which aim to interfere with tumour growth and survival and kill proliferating cells, this class of therapeutic mainly interferes with the molecular functions required for tumour growth and development (Gerber 2008). Cell surface proteins targetable with monoclonal antibodies (mAbs) or derivatives of antibodies represent ideal candidates for tailored treatment in patients. They can be used not only to identify patients that benefit from treatment but also as a target to stop spread of disease. The ideal cell surface protein is one that is expressed with high density on target pathology with low/no expression on normal tissue. Therefore, treating based on cell surface expression of a specific protein could significantly improve efficacy and minimize the on-target side effects of treatments.

Currently most targeted therapies available are either monoclonal antibodies or small-molecule inhibitors (Gerber 2008, Charmsaz, Scott, and Boyd 2017). Monoclonal antibodies available in the clinic for treatment of different malignancies include, bevacizumab targeting VEGF in colorectal cancer and non-small lung cancer (McCormack and Keam 2008, Russo et al. 2017), cetuximab targeting EGFR in head and neck cancer (Specenier and Vermorken 2013) and rituximab targeting CD20 for treatment of Non-Hodgkin's lymphoma (Dotan, Aggarwal, and Smith 2010). Small molecule inhibitors for cancer treatment include imatinib, targeting BCR-ABL, c-KIT and Platelet-derived growth factor receptor (PDGFR), is used for treatment of chronic myeloid leukemia (Sacha 2014) and sorafenib which has been used for treatment of hepatocellular carcinoma (Xie, Wang, and Spechler 2012). In recent years new strategies have been employed to improve the efficacy of targeted therapies. These include the use of monoclonal antibodies conjugated to a payload including cytotoxic drugs, siRNAs and radiolabelled isotopes (Charmsaz and Boyd 2017, Charmsaz, Al-Ejeh, et al. 2017, Charmsaz et al. 2015) with some of these therapeutics already in clinical use.

Breast cancer is a heterogeneous disease with many different therapeutics currently in use or in clinical trial, here we will describe some of the classic and emerging targeted treatments for the disease.

BREAST CANCER

Breast cancer is the most common cancer in women, with more than 2 million new cases diagnosed in 2018. It is a heterogeneous disease, subdivided into several distinct groups based on clinical, histological and molecular phenotype. These include tumour morphology, stage of disease and receptor status, all of which are combined to assess overall patient prognosis.

The primary tumour is initially assessed at the time of biopsy and/or surgical resection. Evaluation of the three main receptors, estrogen receptor (ER), progesterone receptor (PR) and human epidermal growth factor receptor 2 (HER2) are currently the most important biological stratification of breast cancer. In this classification, originally introduced in 1970's, both ER and PR are implicated as prognostic biomarkers (Lemon 1970, Horwitz and McGuire 1975). In 1987 HER2 was identified as both a new prognostic marker and therapeutic target (Slamon et al. 1987). To date these three receptors form the basis of breast cancer molecular subtyping which categorises the tumour into three distinct groups, hormone receptor positive (ER+ and/or PR+), HER2 positive (HER2+), or triple negative (ER- PR- and HER2-).

The standard management of breast cancer across all subtypes includes pre-operative neo-adjuvant therapies to reduce tumour mass, surgical removal of the primary tumour and adjuvant therapy to eliminate residual tumour cells (Hortobagyi 1998). Adjuvant and neo-adjuvant therapy comprise of chemotherapy, radiation therapy and directed therapies administered based on molecular subtype. Here we discuss current and emerging targeted treatments for breast cancer patients (Wahba and El-Hadaad 2015, Lumachi, Santeufemia, and Basso 2015, Loibl and Gianni 2017).

CURRENT TARGETED THERAPIES FOR BREAST CANCER

Breast cancer is one of the first malignancies for which targeted therapies were successfully applied in clinic. Currently two main molecular targets have been identified and widely used, ER, expressed in 70% and HER2 expressed in 15-20%, of invasive breast cancers (Joshi H and MF 2018).

Endocrine Therapy

ER induces a coordinated activation of oncogenic growth pathways which can be targeted by anti-endocrine treatment. Endocrine therapies directed against steroid receptors are the cornerstone of systemic treatment, tumours that express greater than 1% of either ER or PR are considered suitable for treatment (Hammond et al. 2010). Currently, three main classes of endocrine treatment are in clinical use, estrogen receptor modulators (SERM), selective estrogen receptor down-regulators (SERD) and aromatase inhibitors (AI).

SERMs function by competitive inhibition of estrogen, binding to the ER, affecting both genomic and non-genomic receptor activity (Jordan and O'Malley 2007). Tamoxifen is the first class of SERM approved by U.S Food and Drug Administration (FDA) and was initially used for treatment of metastatic breast cancer irrespective of ER status (Pierson and Swann 1991). Since approval several studies and clinical trials have demonstrated long term (up to 10 years) benefit of tamoxifen in ER-positive patients (Cohen et al. 2001). In ER-positive patients the active metabolites of tamoxifen, afimoxifene (4 Hydroxytamoxifen; 4-OHT) and endoxifen function as ER antagonists in breast tissue, inhibiting transcription of estrogen-responsive genes, but can also act as partial agonists in endometrium and bone (Wang et al. 2004). Side effects associated mainly with the agonist properties of tamoxifen include menopausal symptoms, an increased risk of endometrial cancer and thromboembolism (Smith and O'Malley 2004). These side effects in patients have prompted the development of new ER targeted agents with different mechanisms of action.

Fulvestrant is a class of SERD which functions to either down regulate ER or reduce ER stability (Kabos and Borges 2010). Having a higher affinity for ER than tamoxifen, fulvestrant inhibits ER dimerization, can promoting nuclear export and facilitating receptor degradation (Wakeling 2000). Fulvestrant is almost an exclusive ER antagonist and lacks pro-estrogenic properties associated with tamoxifen (Moverare-Skrtic et al. 2014). Treatment of ER-positive patients with fulvestrant has shown similar efficacy compared to other endocrine therapies. New studies, however indicate that higher doses of fulvestrant (500 mg), compared to the original standard lower dose (250 mg) can result in improved progression free survival and is potentially more effective in treatment of advanced ER positive breast cancer (Nathan and Schmid 2017).

AIs are used to treat postmenopausal women or those who have underwent treatment to induce ovarian suppression. AI therapy functions by inhibiting CYP19, the aromatase enzyme required for conversion of androgens to estrogen. There are two main classes of AIs, steroidal and non-steroidal agents (Ma et al. 2015). Exemestane, a steroidal AI, binds covalently and irreversibly to the aromatase enzyme whereas letrozole and anastrozole are non-steroidal AIs that bind non-covalently and reversibly to CYP19 (Santen et al. 2009). Overall AIs have outperformed tamoxifen in clinical trials and are now considered the first-line therapy for postmenopausal ER-positive patients (Chumsri et al. 2011).

Although endocrine therapies have proven successful in the treatment of breast cancer, approximately 30% of patients will develop endocrine resistance and disease recurrence. Trials evaluating the extended use of tamoxifen and AI beyond the standard duration of 5 years identified a modest improvement in outcomes, however these must be weighed against the increased toxicity of extended endocrine therapy regimes (Davies et al. 2013, Goss et al. 2016). New studies indicate that combined endocrine therapy with molecular targeted agents that inhibit pathways leading to development of resistance can result in an improved outcome in ER positive patients. Some of these endocrine therapies combined with targeted agents are currently in clinical trial and under investigation (Jiang, Zheng, and Wang 2013).

HER2 Targeted Therapies

ERBB2 (HER2) is a tyrosine kinase receptor and part of the epidermal growth factor receptor family. Breast cancers in which *ERBB2* gene amplification is present are considered to be HER2-positive (Wolff et al.

2013). Amplification of *ERBB2* was initially described as an adverse prognostic indicator (Slamon et al. 1987) for breast cancer patients. Multiple targeted therapies have been developed for patients with tumours that have *ERBB2* amplification. These include the use of monoclonal antibodies trastuzumab and pertuzumab as well as small molecule inhibitors neratinib and lapatinib. Approximately 20% of all breast cancers are HER2-positive. Prior to the widespread clinical use of HER2 targeted therapies, only 2-5% of patients had long-term survival, however the use of these therapies have significantly improved patient outcome (Wilson et al. 2017).

Trastuzumab targets ERBB2 by binding to the juxtamembrane region of the extracellular domain, limiting intrinsic tyrosine kinase activity and inhibiting downstream signalling pathways. In turn, pertuzumab, a second generation anti-ERBB2 monoclonal antibody targets the receptor dimerization domain and interferes with the mechanism of oncogenic signal generation (Nami, Maadi, and Wang 2018). Neratinib and lapatinib are non-specific small molecule inhibitors of human epidermal growth factor receptor kinases of which ERBB2 is a member. Neratinib is a pan tyrosine kinase inhibitor that functions by interacting with the catalytic domain of members of EGFR family blocking their downstream signalling pathways (Dhillon 2019). Lapatinib is a reversible tyrosine kinase inhibitor that targets both ERBB2 and EGFR and is mainly used in combination with chemotherapeutic agents for the treatment of patients who have previously failed on trastuzumab (Segovia-Mendoza et al. 2015).

In recent decades, the development and efficacy of ERBB2 targeted therapies has been one of the great developments in breast cancer treatment (Piccart-Gebhart et al. 2005, Romond et al. 2005). Resistance and metastasis, however remain a major clinical challenge in this subgroup of patients. New agents targeting ERBB2 receptors, downstream effectors and signalling pathways as well combinational therapies, are currently in development or in clinical trial to improve the outcome of HER2-positive breast cancer patients (Li and Li 2013).

Targeted Therapy for Triple Negative Breast Cancer

Triple negative breast cancer (TNBC) consist of tumours which are ER-negative, PR-negative and HER2-negative and represent 10-20% of all breast cancers. There are currently limited targeted treatment options for this cohort of patients. Systemic chemotherapy together with radiotherapy and surgery are still the mainstay of treatment for TNBC patients. Platinum compounds have shown good efficacy in patients with BRCA1/2 mutations. Other targeted therapies currently under clinical investigation include phosphoinositide 3-kinase (PI3K) inhibitors, MEK inhibitors, anti-androgen therapies, heat shock protein 90 inhibitors and histone deacetylase inhibitors (HSP90-inhitoe) (Bianchini et al. 2016). Some of these agents are already in clinical trial.

Multi-Gene Signature Assays

Recent biological developments in breast cancer including high throughput gene expression profiling have opened new avenues to further stratify breast cancer patients. There are seven molecular tests currently available, OncotypeDX, MammaPrint, Prosigna, EndoPredict, Breast Cancer Index, Mammostrat, and IHC4 (Vieira and Schmitt 2018). Some of these tests are used to identify ER-positive patients who are suitable for endocrine treatment alone or in combination with additional chemotherapy (Harris et al. 2016). There are two main FDA approved products currently in clinical practice for predicting outcome.

Oncotype Dx® is a gene panel molecular assay that uses real-rime quantitative reverse transcription polymerase chain reaction (qRT-PCR) to quantify the expression of a 21 gene set which includes ER, ERBB2 and ER-regulated transcripts as well as genes associated with proliferation to calculate a recurrence score (Harris et al. 2016). The American Society of Clinical Oncology (ASCO) guideline as well as the guideline from American Joint Committee on Cancer (AJCC) indicate that the Oncotype Dx should not be used to guide treatment decisions in HER2-positive or triple negative breast cancer. This gene signature assay is however, used to indicate the best adjuvant systemic therapy in ER-positive, lymph node-negative disease (Siow et al. 2018, Vieira and Schmitt 2018).

Mammaprint® is another gene signature assay using a panel of 70 genes (Cardoso et al. 2016). It is designed to predict the risk of metastasis in early breast cancer patients (Brandao, Ponde, and Piccart-Gebhart 2019) and is now used to classify patients with ER-positive or ER-negative disease into high risk or low risk groups (Wuerstlein et al. 2019).

It is important to note that the development of genomic assays have the potential to significantly improve the quality of life for patients with indolent disease.

TARGETED THERAPIES IN METASTATIC BREAST CANCER

Metastatic breast cancer accounts for the vast majority of breast cancer related deaths. In the last decade there have been significant developments in the area of targeted therapy for advanced disease, including the use of CDK4/6 inhibitors, poly adenosine diphosphate-ribose polymerase (PARP) enzyme inhibitors and vascular endothelial growth factor (VEGF) inhibitors.

Abemaciclib, palbociclib and ribociclib are CDK4/6 inhibitors that have been approved for treatment of hormone receptor positive metastatic breast cancer (Finn et al. 2016, Goetz et al. 2017, Hortobagyi et al. 2016, Sledge et al. 2017, Dickler et al. 2017, Turner et al. 2015, Slamon et al. 2018).

Bevacizumab is a VEGF inhibitor which acts as an anti-angiogenic to block new blood vessel formation. In 2011 the FDA restricted the use of this drug in metastatic breast cancer as clinical results did not demonstrate efficacy or improve patient overall survival (Crown, O'Shaughnessy, and Gullo 2012). Since then a number of studies and clinical trials have been carried out to evaluate the use of this drug in combination with other agents in metastatic breast cancer, as well as in neo-adjuvant and adjuvant settings in early stage disease. New investigations and clinical trials have reported promising outcomes for bevacizumab (Manso et al. 2015, Dank et al. 2014).

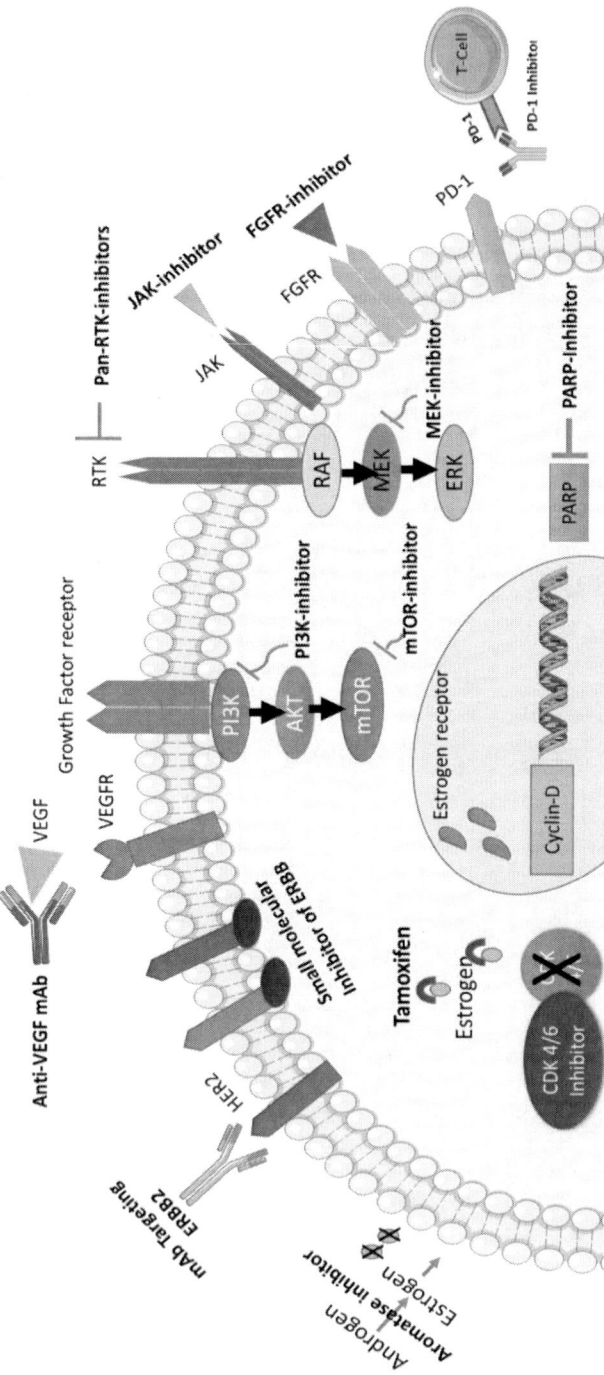

Figure 1. The schematic representation of the targeted therapies for breast cancer treatment.

Table 1. Targeted therapies in clinical practice

Drug	Mechanism of action
Tamoxifen	ER antagonist in breast tissue, inhibiting transcription of estrogen-responsive genes; partial agonist in endometrium and bone.
Fulvestrant	ER antagonist; no agonist activity.
Letrozole, Anastrazole Exemestane	Aromatase inhibitor; inhibits the conversion of peripheral circulating androgens to estrogen.
Trastuzumab	Recombinant humanised monoclonal antibody; targets the extracellular domain of ERBB2.
Pertuzumab	Second generation anti-ERBB2 monoclonal antibody; targets the ERBB2 dimerization domain and interferes with the mechanism of oncogenic signal generation.
Lapatinib	Small molecule inhibitor of human epidermal growth factor receptor (EGFR) kinases of which ERBB2 is a member.
Neratinib	Non-specific small molecule inhibitor of human epidermal growth factor receptor kinases of which ERBB2 is a member.
Bevacizumab	Humanised anti-vascular endothelial growth factor (VEGF) monoclonal antibody.
Everolimus	Mammalian target of rapamycin (mTOR) inhibitor; binds to protein receptor FKBP12; selective for mTORC1 protein complex, results in increased AKT kinase signalling and disrupts cell cycle proteins.
Palbociclib	CDK4/6 inhibitor; inhibits progression through the cell cycle.
Ribociclib	Small molecule CDK4/6 inhibitor; inhibits phosphorylation of Rb via inhibition of CDK4/6-cyclin D complex preventing the cell from passing restriction point R and passing through G1 phase of cell cycle.
Abemaciclib	Selective CDK4/6 inhibitors.
Olaparib	PARP enzyme inhibitor; Interferes with normal cellular DNA damage repair specifically in BRCA deficient cells.

PARP enzyme inhibitors have demonstrated efficacy in patients with deleterious germline BRCA mutations in recent clinical trials (Robson et al. 2017). BRCA1 or BRCA2 germline mutations account for approximately 5% of breast cancers. PARP enzyme inhibitors interfere with normal cellular DNA damage repair, specifically in BRCA deficient cells. PARP enzyme inhibitors, olaparib and talazoparib were recently approved for the treatment ERBB2 negative refractory metastatic breast cancer with deleterious BRCA1/2 germline mutations.

A number of alternative signalling pathways have also been described as a mechanism of endocrine resistance in hormone receptor positive breast cancer. These signalling pathways provide a focus for investigation of several potential targeted therapies. Signalling networks that may be amenable to targeted therapies include ERK signalling pathway, phosphoinositide 3-kinase (PI3K), VEGF and extra cellular matrix members (S100β, ADAM22) (Charmsaz, Hughes, et al. 2017, Bolger and Young 2013), which are currently either under clinical or pre-clinical investigations and will be discussed in the next section. A summary of FDA approved targeted therapeutics for breast cancer patients are provided in Table 1 and Figure 1.

EMERGING THERAPEUTICS
FOR TREATMENT OF BREAST CANCER

Recent advances have led to the development of new effective targeted therapies for all subtypes of breast cancer. Here some of the agents currently undergoing clinical trial for the treatment of breast cancer are discussed.

Targeting RTKs and Downstream Signalling Pathways in Breast Cancer

Overexpression or dysregulation of different members of receptor tyrosine kinases (RTKs), including epidermal growth factor receptors (EGFRs), vascular endothelial growth factor receptors (VEGFRs), Eph family of receptors, insulin-like growth factor receptors (IGFRs), fibroblast growth factor receptors (FGFRs) and hepatocyte growth factor receptor (HGFR/MET), are found in different types of malignancies including breast cancer and are shown to associate with aggressiveness and decreased progression free survival in patients (Cecchi, Rabe, and Bottaro 2011, Minuti and Landi 2015, Charmsaz and Boyd 2017, Boyd, Bartlett, and Lackmann 2014, Charmsaz and Boyd 2013). RTKs also affect disease progression through activation of various downstream signalling pathways including mitogen-activated protein kinase (MAPK), Janus kinase (JAK)/signal transducer and activator of transcription (STAT) and phosphoinositide 3-kinase (PI3K)/Akt pathways (Cecchi, Rabe, and Bottaro 2011).

As RTKs have demonstrated a pivotal role in breast cancer progression and development, they have become a promising target in the treatment and management of the disease. Many drugs targeting RTKs or their downstream signalling pathways are currently in clinical use or under investigation.

Targeting RTKs in Breast Cancer

Pan-Receptor Tyrosine Kinases Inhibitors

Receptor tyrosine kinases (RTKs) are known to be involved in a variety of cellular processes including cell growth, motility, differentiation and metabolism. Their expression and genetic alteration is also associated with development and progression of different malignancies. Drugs targeting RTKs currently under investigation include cabozantinib, lucitanib and nintedanib. Cabozantinib is an orally bioavailable tyrosine kinase inhibitor with potent activity against MET, VEGFR2, AXL and other RTKs (Yakes et al. 2011) and has demonstrated efficacy in xenograft models (Tolaney et al. 2016). Furthermore, clinical trials have shown promising outcomes in patients with heavily pre-treated metastatic breast cancer. Currently, trials evaluating the role of cabozantinib in combination with other agents are ongoing in the metastatic setting (Table 2).

Lucitanib is a potent, orally available small molecule inhibitor of VEGFR1-3, FGFR1-2, and CSF1R with additional anti-angiogenic and anti-tumour activity. Lucitanib targets proliferation in tumour cells with altered FGF/EGFR pathways (Bello et al. 2011, Bello et al. 2013). Clinical trials evaluating the efficacy of this drug have demonstrated promising outcomes in a phase 1 clinical trial and currently two phase 2 trials are under investigation (Perez-Garcia et al. 2018) (Table 2).

Nintedanib is a triple angio-kinase inhibitor that targets VEGFRs, FGFRs and PDGFRs (Reguera-Nunez et al. 2019). It has been used as a combination therapy in phase I and phase II trials with paclitaxel and docetaxel respectively with promising efficacy. Overall, pan RTK inhibitors have shown good potential both as a single agent and in combination with other therapeutics in breast cancer (Table 2).

Novel ERBB Inhibitors

Currently the main ERBB inhibitors in clinical practice target the receptor and as stated previously some of the widely used members are trastuzumab, a monoclonal antibody and lapatinib, a dual EGFR/HER2 kinase inhibitor.

Afatinib is a new selective inhibitor of the ERBB family of tyrosine kinases and functions by binding and forming homodimers and/ or heterodimers with EGFR, HER2, ERBB3 and ERBB4 resulting in inhibition of tyrosine kinase autophosphorylation and downregulation ERBB signalling pathways (Deeks and Keating 2018). Afatinib monotherapy has demonstrated promising clinical outcome, specifically in patients who have progressed on other HER2 inhibitors including trastuzumab. There are currently several new studies looking at afatanib in combination with other agents with results pending publication (Harbeck et al. 2016) (Table 2).

Sapitinib is a novel, equipotent, reversible small-molecule ATP competitive tyrosine kinase inhibitor of EGFR, HER2 and HER3 signalling that has shown anticancer activity in a range of *in vitro* and *in vivo* preclinical models. Sapitinib is significantly more potent against EGFR, HER2 and HER3 signalling *in vitro,* and has greater anti-tumour efficacy in xenograft models compared to other EGFR inhibitors including lapatinib (Hickinson et al. 2010). Combination therapy with AKT-inhibitors (AZD5363) have demonstrated efficacy in the clinic for the treatment of HER2 positive breast cancer (Crafter et al. 2015). Other studies are currently evaluating the efficacy of sapitinib in combination with other agents (Table 2).

FGFR Inhibitors

The fibroblast growth factor receptor (FGFR) family consists of five transmembrane receptors, where all except one has tyrosine kinase activity. Recently, a number of studies have focused on determining the role of FGFR signalling in the progression of different malignancies. These studies have demonstrated a role for this family of proteins in cancer cell proliferation, angiogenesis and survival. The role of FGFR in cancer progression makes this family of protein an excellent target for cancer therapy (Perez-Garcia et al. 2018). Aberrant genomic alterations in FGFR signalling pathways have been detected in breast cancer, a number of agents that target FGFR signalling are currently being evaluated in clinical and pre-clinical studies, including AZD4547 and erdafitinib.

Table 2. RTK inhibitors in clinical trial

	Pan-RTK inhibitor
Cabozantinib	Cabozantinib is an orally bioavailable tyrosine kinase inhibitor with potent activity against MET, VEGFR2, AXL and other RTKs. A Phase 2 and Biomarker Study (NCT01738438) of cabozantinib in metastatic triple-negative breast cancer patients evaluated the safety, efficacy and response to treatment. From 35 patients who initiated protocol therapy, 3 (9%) achieved a partial response (PR) and 9 (27%) had stable disease (SD) for at least 15 weeks. The clinical benefit rate was calculated (PR+SC) to be 34% with a median progression free survival of two months. While analyses of circulating biomarkers showed that cabozantinib induces systemic changes consistent with activation of the immune system and antiangiogenic activity, it was concluded that the primary endpoint in pre-treated metastatic triple-negative breast cancer was not met. Two phase 2 randomised trials are currently active and no longer recruiting. These studies are evaluating the effect of cabozantinib +/- trastuzumab In breast cancer patients with brain metastases (NCT02260531) and the effect of cabozantinib treatment in women with metastatic hormone receptor positive breast cancer (NCT01441947).
Lucitanib	Lucitanib is a potent, orally available small molecule inhibitor of VEGFR1-3, FGFR1-2 and CSF1R. A phase 2 study of lucitanib (NCT02202746) was recently completed with results pending publication. In this trial lucitanib safety and efficacy in the treatment of patients with FGF aberrant metastatic breast cancer, as well as in the treatment of patients with biomarker negative (FGF non-aberrant) metastatic breast cancer, was assessed. Another recently completed trial is FINESSE (NCT02053636) - a phase 2 trial testing oral administration of lucitanib in patients with FGFR1-amplified or non-amplified estrogen receptor-positive metastatic breast cancer. Results from this study are yet to be published.
Nintedanib	Nintedanib is a triple angiokinase inhibitor which targets VEGFRs, FGFRs and PDGFRs. Nintedanib in combination with paclitaxel has been evaluated in a phase 1 clinical trial and in early HER2-negative breast cancer patients showing promising antitumor efficacy (Quintela-Fandino et al. 2014).

Table 2. (Continued)

Nintedanib (Continued)	**Pan-RTK inhibitor**	
	Currently, there are two clinical trials testing nintedanib in combination with paclitaxel (NCT01484080, phase I/II) and docetaxel (NCT01658462, phase II) in early and metastatic or locally recurrent HER2-negative breast cancer, respectively.	
Afatinib	**Novel ERBB inhibitors**	
	Afatinib is an anilinoquinazoline-derived, oral small-molecule ERBB family blocker that inhibits EGFR, HER2, ERBB4 and ERBB3 trans-phosphorylation. Afatinib is currently approved in several countries for the treatment of patients with non–small cell lung cancer with *EGFR* tyrosine kinase mutations. It has demonstrated clinical activity in advanced solid tumours and in trastuzumab-refractory HER2 positive breast cancer (Ring et al. 2015).	
Sapitinib - AZD8931	**Novel ERBB inhibitors**	
	Sapitinib is a novel, equipotent, reversible small-molecule ATP competitive tyrosine kinase inhibitor of EGFR, HER2 and HER3 signalling. A phase 2 randomized placebo-controlled study (NCT00900627) of AZD8931 plus paclitaxel vs paclitaxel alone in patients with low HER2 expressing advanced breast cancer (THYME), demonstrated no significant difference in progression free survival with AZD8931 + paclitaxel vs paclitaxel alone. This lack of treatment efficacy was consistent across subgroups. A significant increase in overall response rate (59% vs 41%) and in mean percentage change of tumour size at week 8 vs baseline (difference, -9.9%,) was observed with AZD8931 + paclitaxel vs paclitaxel alone. Overall survival data was immature (52 [27%] pts had died), with no difference between treatment arms. The THYME study did not meet its primary objective of prolonging progression free survival when AZD8931 was added to weekly paclitaxel in patients with low HER2-expressing advanced breast cancer.	
AZD4547	**FGFR inhibitor**	
	AZD4547 is a multi-kinase inhibitor which targets FGFRs. A phase 2 randomized trial (NCT02299999) compares targeted kinase inhibitor treatment, including AZD4547 (administered according to the identified molecular anomalies of the tumour) to maintenance chemotherapy (Sub-study 1). In this clinical trial patients without actionable genomic alterations and patients who are not eligible to enter Sub-study 1 will enter Sub-study 2, where they compare treatment with immunotherapy to standard maintenance chemotherapy.	
Erdafitinib	Erdafitinib is a pan-FGFR inhibitor. A Phase 1b trial (NCT03238196) of fulvestrant, palbociclib (CDK4/6 Inhibitor) and erdafitinib (JNJ- 42756493, Pan-FGFR Tyrosine Kinase Inhibitor) in ER+/HER2-/FGFR-amplified metastatic breast cancer is currently recruiting. The primary objective is to determine the safety and tolerability of fulvestrant, palbociclib and erdafitinib treatment, secondary outcomes will assess the anti-tumour effect of the drug combination in this patient population.	

AZD4547 is a multi-kinase inhibitor targeting FGFRs, CSF1R, VEGFR2 and others. FGFR inhibitors elicit antitumor effects directly on cancer cells, as well as indirectly through the blockade of paracrine signalling. The dual inhibition of FGF and CSF1 or VEGF signalling has been reported to enhance the antitumor effects by targeting immune evasion and angiogenesis in the tumour microenvironment (Perez-Garcia et al. 2018). The phase I/II clinical trials evaluating the effect of AZD4547 have demonstrated anti-cancer activity and tumour regression in breast cancer patients (Zhao et al. 2017), confirming the potential of this drug as a therapeutic agent. Currently new clinical trials are evaluating the effect of AZD4547 as a combinational therapy (Table 2).

Erdafitinib is a pan-FGFR inhibitor which, unlike other FGFR inhibitors (lucitanib), does not target VEGFR2 (Tabernero et al. 2015). Combined treatment erdafitinib is currently in clinical trial (Table 2).

Targeting Downstream Signalling Pathways in Breast Cancer

PI3K Inhibitors

Phosphatidylinositol-3 kinase (PI3K) and mitogen-activated protein kinase (MAPK) pathways are pro-survival and pro-growth, and are associated with the oncogenesis of several cancers including breast. PI3K pathway activation is involved in the development of resistance in both ER positive and HER2 amplified disease (Keegan et al. 2018) and represents a potential therapeutic target for resistant breast cancer. Several agents targeting the PI3K pathway have been developed and are currently in clinical trial. Buparlisib, a potent and specific pan inhibitor of the PI3K family, inhibits tumour cell growth and survival, and is currently under investigation in different types of malignancies including breast cancer. The results from buparlisib monotherapy however have not been promising. It is currently being evaluated in ongoing clinical trials in combination with other agents including letrozole in post-menopausal women with advanced hormone receptor positive breast cancer (Maira et al. 2012). Pictilisib, another oral PI3K inhibitor, is currently undergoing clinical trial in hormone receptor positive, AI resistant patients (Lee, Loh, and Yap 2015). These studies suggest that while PI3K inhibitors as monotherapy have not been successful, ongoing trials are providing promising data for combination strategies to maximize therapeutic outcomes.

**Table 3. RTKs downstream signalling pathway inhibitors
in clinical trial**

	PI3K inhibitor
Pictilisib (GDC0941)	Pictilisib is an oral PI3K inhibitor. A phase 1b clinical trial (NCT00960960) of pictilisib (GDC-0941) with paclitaxel, with or without bevacizumab or trastuzumab, or in combination with letrozole is currently under investigation in patients with locally recurrent or metastatic breast cancer (Schoffski et al. 2018).
Buparlisib (BKM120)	Buparlisib is a potent and specific pan inhibitor of PI3K family. Buparlisib (BKM120) is currently under investigation in metastatic breast cancer. A phase I study was designed to establish the maximum tolerated dose (NCT01300962).
	MAPK inhibitor
Selume-tinib	Selumetinib, a targeted small-molecule drug, is an allosteric, ATP non-competitive inhibitor of MEK1/2. A phase 2 randomized trial (NCT02299999) compares a targeted MAPK inhibitor treatment including selumetinib, (administered according to the identified molecular anomalies of the tumour) to maintenance chemotherapy (Sub-study 1). In this clinical trial patients without actionable genomic alterations and patients who are not eligible to enter Sub-study 1 will enter Sub-study 2, where they compare treatment with immunotherapy to standard maintenance chemotherapy.
	JAK inhibitor
Ruxolitinib	Ruxolitinib is an orally bioavailable receptor tyrosine kinase inhibitor which targets JAK1 and JAK2. Currently a number of clinical trials are evaluating ruxolitinib as part of a combination therapy in breast cancer. A Phase 2 study of ruxolitinib (INCB018424) with preoperative chemotherapy for triple negative inflammatory breast cancer (NCT02876302) is ongoing and recruiting patients. In this trial the investigators are evaluating ruxolitinib in combination with paclitaxel followed by the standard chemotherapy. The study is due for completion in 2024.
	MET inhibitor
Tivantinib	Tivantinib is a selective, oral, small molecule inhibitor of the MET receptor tyrosine kinases. A Phase 2 study (NCT01575522) of tivantinib patients with metastatic triple-negative breast cancer, who had received 1 to 3 prior lines of chemotherapy in the metastatic setting were enrolled. Treatment consisted of twice-daily oral dosing of tivantinib, during a 21-day cycle. 22 patients were enrolled. The overall response rate was 5% and 6-month progression-free survival was increased with one patient achieving partial response. However this study failed to meet statistical targets for efficacy, but the drug was well tolerated (Tolaney et al. 2015). Treatment with tivantinib and other agents are currently in clinical trial (NCT01178411).

Dysregulation of MAPK pathways is associated with disease progression. MAPK pathway acts through a cascade of phosphorylation pathways and can ultimately activate transcription factors including ER (Gee et al. 2001). Targeting MAPK pathways has demonstrated promising outcomes. Selumetinib, a targeted small-molecule drug which is an allosteric, ATP non-competitive inhibitor of MEK1/2, has shown activity against several malignant tumours (Maiello et al. 2015). Selumetinib as a therapeutic agent in combination with other therapies is currently in clinical trial (Table 3).

JAK Inhibitors

Janus tyrosine kinase (JAK) and signal transducers and activators of transcription (STATs) have a key role in the development, proliferation, differentiation and survival of various types of cancer cells including breast cancer. JAKs are a family of cytoplasmic receptor associated with protein tyrosine kinases required for cytokine signalling. Ruxolitinib is a receptor tyrosine kinase inhibitor which targets JAK1 and JAK2. The JAK/STAT pathway is involved in normal mammary gland development. In breast cancer this pathway is hijacked resulting in over activation leading to tumorigenesis (Caffarel et al. 2012, Quintas-Cardama et al. 2011, Chang et al. 2013). Clinical trials targeting JAK/STAT pathway are described in Table 3.

MET Inhibitors

Mesenchymal-epithelial transition factor (MET) is a member of the receptor tyrosine kinase family of proteins encoded by MET proto-oncogene. Hepatocyte growth factor (HGF) is expressed on the surface of epithelial cells and is a ligand for MET. MET is normally expressed at a very low level on cells, however expression and activation of this protein is known to be elevated in a number of solid tumours including breast cancer. It has been shown that MET/HGF pathways have an important role in tumour proliferation and invasion (Zhang et al. 2017). In recent years clinical trials have been ongoing in breast cancer to evaluate the efficacy of MET inhibition. Tivantinib is a selective, oral, small molecule inhibitor of the MET receptor tyrosine kinases currently being evaluated in TNBC (Zhang et al. 2017) (Puzanov et al. 2015). Early results indicate that tivantinib is well tolerated with low toxicity, however it has poor clinical efficacy when used as monotherapy to treat unselected, metastatic TNBC (Table 3).

Androgen Receptor Antagonist/Modulators in Breast Cancer Treatment

Androgen receptor (AR) has been detected in all stages of breast cancer and is present at varying degrees among different molecular subtypes.

Table 4. Androgen receptor antagonist/modulator as potential therapeutic target for treatment of breast cancer

	Androgen receptor antagonist/modulator
Bicalutamide	Bicalutamide is a peripherally selective non-steroidal anti-androgen that has demonstrated efficacy in the treatment of triple negative breast cancer patients with metastatic disease. Bicalutamide, in combination with palbociclib, is currently the focus of an ongoing trial in the treatment of metastatic breast cancer (NCT02605486).
Enzalutamide	Enzalutamide is a peripherally selective non-steroidal anti-androgen that has demonstrated efficacy in the treatment of triple negative breast cancer patients with metastatic disease. Current clinical trials evaluating the efficacy of enzalutamide in combination with other therapeutics are ongoing. These include alpelisib in combination with enzalutamide for treatment of AR+ve/PTEN+ve metastatic breast cancer (NCT03207529), trastuzumab in combination with enzalutamide for treatment of HER2 positive breast cancer (NCT02091960), anastrazole/exemestane/fulvestrant in combination with enzalutamide in ER+ve/AR+ve breast cancer (NCT01597193) and paclitaxel in combination with enzalutamide in AR+ve triple negative breast cancer (NCT02689427).
Enobosarm (GTx-24)	Enobosarm is an investigational selective androgen receptor modulator. Currently the efficacy of enobasarm in combination with the anti-PD-1 immunotherapy drug pembrolizumab in AR+ve/ER-ve/PR+ve recurrent advanced breast cancer is underway (NCT02971761) (Bleach and McIlroy 2018).

Studies have shown that AR is expressed widely on breast cancer cells (Iacopetta, Rechoum, and Fuqua 2012), suggesting its potential as a biomarker and/or therapeutic target. Current clinical trials are evaluating AR antagonists as well as AR modulators and have reported the overall clinical benefit of targeting AR as a monotherapy. Studies examining AR combination therapies are currently ongoing with no outcome data reported to date. However, early studies suggest that AR therapy could greatly benefit from a robust biomarker to stratify patients who may benefit most from AR treatment (Bleach and McIlroy 2018). AR inhibitors include bicalutamide and enzalutamide, both of which are selective non-steroidal anti-androgens. AR modulators include enobosarm which is a selective androgen receptor modulator currently in clinical trial. Table 4 lists androgen receptor antagonists/modulators that are in clinical trial (Table 4).

PARP Inhibitors in Breast Cancer Treatment

Poly (ADP-ribose) polymerase inhibitors (PARPi's) primarily target PARP1, an enzyme involved in the repair of single-stranded DNA breaks (Audeh 2014). PARP inhibition is emerging as a promising treatment strategy in patients with a BRCA1/2 mutation. As mentioned a functioning PARP system repairs DNA double-strand breaks, however when inhibited, cells must rely on homologous recombination for repair, therefore requiring functional BRCA1 and BRCA2. The combined inhibition of PARP1 and loss of homologous recombination repair, is termed "synthetic lethality". Synthetic lethality causes an accumulation of double-strand breaks and ultimately leads to cell death (Zimmer et al. 2018, Lee and Djamgoz 2018). Combination regimens including PARP inhibitors sensitize BRCA mutant patients, improving PARP inhibitor efficacy and selectivity (Audeh 2014). A number of PARP inhibitors including rucaparib, niraparib, talazoparib, veliparib and fluzoparib are under investigation for the treatment of breast cancer. Some of the clinical trials with PARP inhibitors are listed in Table 5.

Table 5. PARP inhibitors as potential therapeutic target
for treatment of breast cancer

	PARP inhibitors
Rucaparib	Rucaparib is a potent small-molecule inhibitor of PARP-1, PARP-2 and PARP-3 which has been developed for the treatment of ovarian cancer and other tumour types associated with homologous recombination deficiency (HRD), including BRCA1 and BRCA2 mutations (Schreiber et al. 2006). The phase 2 'RUBY' trial (NCT02505048) assessing rucaparib efficacy in patients with a 'BRCAness' phenotype. Primary clinical outcome of either complete response (CR), partial response (PR) or stable disease (SD) lasting for at least 16 weeks, according to RECIST criteria, will be assessed. The active phase 2 trial (NCT01074970) is currently evaluating 2-year disease-free survival in patients treated with single agent cisplatin or cisplatin in combination with rucaparib following preoperative chemotherapy.
Niraparib	Niraparib is an FDA-approved PARP inhibitor used in the treatment of unselected platinum-sensitive recurrent ovarian cancer patients (Scott 2017). TOPACIO trial phase 1 results demonstrated good tolerability for niraparib when used in combination with pembrolizumab for treatment of patients with metastatic TNBC. Phase 2 of this trial is currently active (Zimmer et al. 2018).
Talazoparib	Talazoparib is an investigational PARP inhibitor which has exhibited selectivity towards tumour cells with BRCA1/2 or PTEN gene defects *in vitro* with greater potency than other PARP1/2 inhibitors (Shen et al. 2013). A phase 1 study had demonstrated the efficacy of talazoparib as a single agent (1.0 mg/day) in breast cancer patients (n = 14) with a deleterious BRCA1/2 mutation, where 50% of the patients achieved overall response rate and was well tolerated. The phase 2 study ABRAZO (two stage, two cohort) assessed talazoparib in germline BRCA mutation carriers with response to prior platinum with no progression on or within 8 weeks of the last platinum dose (cohort 1, n = 49) or ≥3 platinum-free cytotoxic regimens (cohort 2, n = 35) for advanced breast cancer. Talazoparib was more effective in cohort 1 [ORR: 21%; duration of response (DOR): 5.8 months; CBR: 38%; PFS: 4.0 months] than in Cohort 2 (ORR: 37%; DOR: 3.8 months; CBR: 66%; PFS: 5.6 months) (Turner et al. 2019). In a subsequent phase 3 study (n = 287), Talazoparib (1.0 mg/day) was compared with the physician's choice therapy (PCT = capecitabine, eribulin, gemcitabine or vinorelbine in continuous 21-day cycles) in patients with advanced breast cancer and a gBRCA1/2 mutation. Progression free survival was significantly longer in talazoparib treated versus the PCT treated group (8.6 versus 5.6 months) without any significant difference observed in terms of overall survival (22.3 versus 19.5 months)(Litton et al. 2018).
Veliparib	Veliparib is an oral PARP-1 and PARP-2 inhibitor which can pass the blood brain barrier and exerts its mechanism of action by potentiating the cytotoxicity of DNA-damaging agents (Murai et al. 2014).

	PARP inhibitors
Veliparib (continued)	An initial phase I study on 35 patients with various advanced solid tumours including breast cancer was conducted to determine the dosing, pharmacokinetics and pharmacodynamics of veliparib in combination with the topoisomerase inhibitor irinotecan (LoRusso et al. 2016). In another phase I study, the efficacy of veliparib was tested in combination with carboplatin (*n* = 27) or as a single agent (*n* = 44) in patients with *gBRCA1/2* metastatic breast cancer. A progression free survival of 8.7 months and an overall survival of 18.8 months in the veliparib-carboplatin combination group was observed compared with veliparib alone (progression free survival and overall survival of 14.4 months and 5.2 months, respectively) (Somlo et al. 2017). A phase 2 study examining veliparib with carboplatin/paclitaxel (VCP), with temozolomide (VT) or placebo plus carboplatin/paclitaxel (PCP) was performed. Of 290 randomized patients, 284 were BRCA+. VCP showed higher clinical benefits compared to PCP where the median progression free survival was 14.1 and 12.3 months, median overall survival of 28.3 and 25.9 months, and overall response rate of 77.8% and 61.3%, respectively. The VT group exhibited worse outcome with median progression free survival 7.4 months, median overall survival 19.1 months and overall response rate of 28.6% (Han et al. 2018).
Fluzoparib	Fluzoparib is an oral potent PARP-1 and PARP-2 inhibitor. A Multi-centre Phase I Study (NCT03075462) is currently recruiting patients which will assess fluzoparib given in combination with apatinib in ovarian and breast cancer patients.

Immunotherapy in Breast Cancer

Cancer immunology is a newly emerging treatment strategy focusing on altering and regulating the immune evasion exhibited by tumour cells (de la Cruz-Merino et al. 2017). However its role in breast cancer is not yet fully understood, with many uncertainties yet to be addressed (de la Cruz-Merino et al. 2017). However, it is known that the host immune response plays a crucial role in breast cancer with many ongoing clinical trials exploring the activity of immune checkpoint modulators in treatment, both in the advanced and neoadjuvant setting. Cancer cells escape the immune response by upregulating immunosuppressive ligands. These ligands may bind to inhibitory co-receptors on the surface of cytotoxic cells and contribute to the shutting down of the anti-tumour immune responses. The main co-inhibitory receptors that have been identified and targeted to date are CTLA-4, PD-1, LAG-3, TIM-3, BTLA, TIGIT and VISTA (PMID: 28629632).

Table 6. Immunotherapy as potential therapeutic target for treatment of breast cancer

	Immunotherapy
Nivolumab	Nivolumab is a monoclonal antibody against programmed cell death (PD-1). PD-1, a receptor belonging to the CD28 protein family, is expressed on many immune cells including antigen presenting cells (APCs), activated lymphocytes and peripheral non-lymphoid cells (Ohaegbulam et al. 2015). PD-L1 ligand expression is positively associated with AKT phosphorylation in both luminal and non-luminal cancers indicating cross-talk with PI3k/AKT pathway at a functional level (Ravelli et al. 2017) (Tsang et al. 2017). The role of PD-1/PD-L1 in different breast cancer subtypes and the benefit of immunotherapy inhibiting this interaction requires further investigation and has been discussed in various reviews (Bertucci and Goncalves 2017, Solinas et al. 2017, Hartkopf et al. 2016). A current phase 2 clinical trial (NCT03789110) is investigating the effect of nivolumab in combination with Ipilimumab (anti-CTLA-4) for the treatment of hyper-mutated HER2 negative breast cancer.
Pembroli-zumab	Pembrolizumab, an antibody against programmed cell death (PD-1), is currently being evaluated for the treatment of different cancers and has been approved in some cases. Pembrolizumab has been investigated as a single agent in recurrent or metastatic triple negative breast cancer. KEYNOTE-012 was a multicohort phase 1b study in which TNBC patients with PD-L1 positive tumours were recruited. The results showed anti-tumour activity (18.5% objective response rate) of pembrolizumab in 111 previously treated patients with advanced TNBC (Nanda et al. 2016). A two part trial of pembrolizumab monotherapy NCT02447003 is currently ongoing investigating the effect of this agent in metastatic TNBC.
Atezoli-zumab	Atezolizumab is an anti PD-L1 antibody which has been investigated as a therapeutic agent in metastatic TNBC. As part of a phase 1 multi-cohort study, atezolizumab was used in heavily pre-treated patients with both PD-L1 positive and PD-L1 negative status (NCT01375842). The phase 1a study demonstrated anti-tumour activity (33% 24-week PFS rate and 24% OR rate) of atezolizumab in 21 previously treated patients with advanced PD-L1-positive triple negative breast cancer. The data was presented at the 2015 AACR meeting.
Avelumab	Avelumab is an anti PD-L1 antibody which was evaluated in JAVELIN, a phase 1b trial in which patients with anthracycline- and taxane- pre-treated metastatic breast cancer of any HER2/ER/PR phenotypes, without preselection for PD-L1 status, were treated with Avelumab (NCT01772004). Phase 1b study of avelumab was conducted in 168 previously treated patients with advanced breast cancers. An 8.8% lower response rate in 57 avelumab treated patients was observed when compared to patients treated with atezolizumab and pembrolizumab (patients had PD-L1 positive tumours). In 9 TNBC patients containing a "hot spot' of PD-L1-positive immune cells within the tumour, the response rate was as high as 44% (Dirix et al. 2018).

Breast cancer can hijack this immunomodulatory mechanism by upregulating these co-inhibitory receptors on the cancer cell surface (Monnot and Romero 2018). It is this understanding of the role of immune checkpoints that has provided the rationale for blocking either the ligands or the receptors using specific monoclonal antibodies based treatments. The first randomized phase III human trial showing the efficacy of such immune checkpoint blockade was published in 2010 and studied the use of a humanized αCTLA-4 antibody called ipilimumab for the treatment of metastatic melanoma. The trial demonstrated that treatment improved both overall survival and progression-free survival. To date in breast cancer, the most promising results have been demonstrated in triple-negative disease. This subgroup is considered to be the most immunogenic due to its higher genomic instability and mutational burden which is responsible for a higher propensity to generate neoantigens, potentially recognized as "non-self" by the adaptive immune system (Monnot and Romero 2018). Thus the immunomodulatory pathway has emerged as a central target for future breast cancer treatments (Hu, Huang, and Fan 2017). Current clinical trials are listed in Table 6.

CONCLUSION

Advances in molecular and genomic biology in recent years have revolutionised the field of breast cancer targeted therapy. Increased drug specificity leads to extended disease free survival with reduced side-effects, ultimately enhancing patient quality of life. Targeting key survival pathways however can result in functional redundancy and may require a combined therapeutic strategy to overcome subsequent targeted drug resistance. Current and future clinical trials will provide valuable information regarding the ideal personalised management strategy to optimise the potential of these new powerful drugs.

REFERENCES

Audeh, M. W. 2014. "Novel treatment strategies in triple-negative breast cancer: specific role of poly(adenosine diphosphate-ribose) polymerase inhibition." *Pharmgenomics Pers Med* 7:307-16. doi: 10.2147/ PGPM.S39765.

Bello, E., G. Colella, V. Scarlato, P. Oliva, A. Berndt, G. Valbusa, S. C. Serra, M. D'Incalci, E. Cavalletti, R. Giavazzi, G. Damia, and G. Camboni. 2011. "E-3810 is a potent dual inhibitor of VEGFR and FGFR that exerts antitumor activity in multiple preclinical models." *Cancer Res* 71 (4):1396-405. doi: 10.1158/0008-5472.CAN-10-2700.

Bello, E., G. Taraboletti, G. Colella, M. Zucchetti, D. Forestieri, S. A. Licandro, A. Berndt, P. Richter, M. D'Incalci, E. Cavalletti, R. Giavazzi, G. Camboni, and G. Damia. 2013. "The tyrosine kinase inhibitor E-3810 combined with paclitaxel inhibits the growth of advanced-stage triple-negative breast cancer xenografts." *Mol Cancer Ther* 12 (2):131-40. doi: 10.1158/1535-7163.MCT-12-0275-T.

Bertucci, F., and A. Goncalves. 2017. "Immunotherapy in Breast Cancer: the Emerging Role of PD-1 and PD-L1." *Curr Oncol Rep* 19 (10):64. doi: 10.1007/s11912-017-0627-0.

Bianchini, G., J. M. Balko, I. A. Mayer, M. E. Sanders, and L. Gianni. 2016. "Triple-negative breast cancer: challenges and opportunities of a heterogeneous disease." *Nat Rev Clin Oncol* 13 (11):674-690. doi: 10.1038/nrclinonc.2016.66.

Bleach, R., and M. McIlroy. 2018. "The Divergent Function of Androgen Receptor in Breast Cancer; Analysis of Steroid Mediators and Tumor Intracrinology." *Front Endocrinol (Lausanne)* 9:594. doi: 10.3389/ fendo.2018.00594.

Bolger, J. C., and L. S. Young. 2013. "ADAM22 as a prognostic and therapeutic drug target in the treatment of endocrine-resistant breast cancer." *Vitam Horm* 93:307-21. doi: 10.1016/B978-0-12-416673-8.00014-9.

Boyd, A. W., P. F. Bartlett, and M. Lackmann. 2014. "Therapeutic targeting of EPH receptors and their ligands." *Nat Rev Drug Discov* 13 (1):39-62. doi: 10.1038/nrd4175.

Brandao, M., N. Ponde, and M. Piccart-Gebhart. 2019. "Mammaprint: a comprehensive review." *Future Oncol* 15 (2):207-224. doi: 10.2217/fon-2018-0221.

Caffarel, M. M., R. Zaragoza, S. Pensa, J. Li, A. R. Green, and C. J. Watson. 2012. "Constitutive activation of JAK2 in mammary epithelium elevates Stat5 signalling, promotes alveologenesis and resistance to cell death, and contributes to tumourigenesis." *Cell Death Differ* 19 (3):511-22. doi: 10.1038/cdd.2011.122.

Cardoso, F., L. J. van't Veer, J. Bogaerts, L. Slaets, G. Viale, S. Delaloge, J. Y. Pierga, E. Brain, S. Causeret, M. DeLorenzi, A. M. Glas, V. Golfinopoulos, T. Goulioti, S. Knox, E. Matos, B. Meulemans, P. A. Neijenhuis, U. Nitz, R. Passalacqua, P. Ravdin, I. T. Rubio, M. Saghatchian, T. J. Smilde, C. Sotiriou, L. Stork, C. Straehle, G. Thomas, A. M. Thompson, J. M. van der Hoeven, P. Vuylsteke, R. Bernards, K. Tryfonidis, E. Rutgers, M. Piccart, and Mindact Investigators. 2016. "70-Gene Signature as an Aid to Treatment Decisions in Early-Stage Breast Cancer." *N Engl J Med* 375 (8):717-29. doi: 10.1056/NEJMoa1602253.

Cecchi, F., D. C. Rabe, and D. P. Bottaro. 2011. "The Hepatocyte Growth Factor Receptor: Structure, Function and Pharmacological Targeting in Cancer." *Curr Signal Transduct Ther* 6 (2):146-151. doi: 10.2174/157436211795659955.

Chang, Q., E. Bournazou, P. Sansone, M. Berishaj, S. P. Gao, L. Daly, J. Wels, T. Theilen, S. Granitto, X. Zhang, J. Cotari, M. L. Alpaugh, E. de Stanchina, K. Manova, M. Li, M. Bonafe, C. Ceccarelli, M. Taffurelli, D. Santini, G. Altan-Bonnet, R. Kaplan, L. Norton, N. Nishimoto, D. Huszar, D. Lyden, and J. Bromberg. 2013. "The IL-6/JAK/Stat3 feed-forward loop drives tumorigenesis and metastasis." *Neoplasia* 15 (7):848-62.

Charmsaz, S., F. Al-Ejeh, T. M. Yeadon, K. J. Miller, F. M. Smith, B. W. Stringer, A. S. Moore, F. T. Lee, L. T. Cooper, C. Stylianou, G. T.

Yarranton, J. Woronicz, A. M. Scott, M. Lackmann, and A. W. Boyd. 2017. "EphA3 as a target for antibody immunotherapy in acute lymphoblastic leukemia." *Leukemia* 31 (8):1779-1787. doi: 10.1038/leu.2016.371.

Charmsaz, S., K. Beckett, F. M. Smith, C. Bruedigam, A. S. Moore, F. Al-Ejeh, S. W. Lane, and A. W. Boyd. 2015. "EphA2 Is a Therapy Target in EphA2-Positive Leukemias but Is Not Essential for Normal Hematopoiesis or Leukemia." *PLoS One* 10 (6):e0130692. doi: 10.1371/journal.pone.0130692.

Charmsaz, S., and A. W. Boyd. 2013. "Expression and function of the Eph receptor family in leukemia and hematopoietic malignancies: prospects for targeted therapies." *Journal of Leukemia*:1-10. doi: 10.4172/2329-6917.1000107.

Charmsaz, S., and A. W. Boyd. 2017. "Eph receptors as oncotargets." *Oncotarget* 8 (47):81727-81728. doi: 10.18632/oncotarget.21045.

Charmsaz, S., E. Hughes, F. T. Bane, P. Tibbitts, M. McIlroy, C. Byrne, S. Cocchiglia, J. McBryan, B. T. Hennessy, R. M. Dwyer, M. J. Kerin, A. D. Hill, and L. S. Young. 2017. "S100beta as a serum marker in endocrine resistant breast cancer." *BMC Med* 15 (1):79. doi: 10.1186/s12916-017-0836-2.

Charmsaz, S., A. M. Scott, and A. W. Boyd. 2017. "Targeted therapies in hematological malignancies using therapeutic monoclonal antibodies against Eph family receptors." *Exp Hematol* 54:31-39. doi: 10.1016/j.exphem.2017.07.003.

Chumsri, S., T. Howes, T. Bao, G. Sabnis, and A. Brodie. 2011. "Aromatase, aromatase inhibitors, and breast cancer." *J Steroid Biochem Mol Biol* 125 (1-2):13-22. doi: 10.1016/j.jsbmb.2011.02.001.

Cohen, M. H., S. Hirschfeld, S. Flamm Honig, A. Ibrahim, J. R. Johnson, J. J. O'Leary, R. M. White, G. A. Williams, and R. Pazdur. 2001. "Drug approval summaries: arsenic trioxide, tamoxifen citrate, anastrazole, paclitaxel, bexarotene." *Oncologist* 6 (1):4-11.

Crafter, C., J. P. Vincent, E. Tang, P. Dudley, N. H. James, T. Klinowska, and B. R. Davies. 2015. "Combining AZD8931, a novel EGFR/HER2/HER3 signalling inhibitor, with AZD5363 limits AKT inhibitor induced feedback and enhances antitumour efficacy in HER2-amplified breast cancer models." *Int J Oncol* 47 (2):446-54. doi: 10.3892/ijo.2015.3062.

Crown, J., J. O'Shaughnessy, and G. Gullo. 2012. "Emerging targeted therapies in triple-negative breast cancer." *Ann Oncol* 23 Suppl 6:vi56-65. doi: 10.1093/annonc/mds196.

Dank, M., L. Budi, B. Piko, L. Mangel, J. Erfan, J. Cseh, A. Ruzsa, and L. Landherr. 2014. "First-line bevacizumab-paclitaxel in 220 patients with metastatic breast cancer: results from the AVAREG study." *Anticancer Res* 34 (3):1275-80.

Davies, C., H. Pan, J. Godwin, R. Gray, R. Arriagada, V. Raina, M. Abraham, V. H. Medeiros Alencar, A. Badran, X. Bonfill, J. Bradbury, M. Clarke, R. Collins, S. R. Davis, A. Delmestri, J. F. Forbes, P. Haddad, M. F. Hou, M. Inbar, H. Khaled, J. Kielanowska, W. H. Kwan, B. S. Mathew, I. Mittra, B. Muller, A. Nicolucci, O. Peralta, F. Pernas, L. Petruzelka, T. Pienkowski, R. Radhika, B. Rajan, M. T. Rubach, S. Tort, G. Urrutia, M. Valentini, Y. Wang, R. Peto, and Group Adjuvant Tamoxifen: Longer Against Shorter Collaborative. 2013. "Long-term effects of continuing adjuvant tamoxifen to 10 years versus stopping at 5 years after diagnosis of oestrogen receptor-positive breast cancer: ATLAS, a randomised trial." *Lancet* 381 (9869):805-16. doi: 10.1016/S0140-6736(12)61963-1.

de la Cruz-Merino, L., M. Chiesa, R. Caballero, F. Rojo, N. Palazon, F. H. Carrasco, and V. Sanchez-Margalet. 2017. "Breast Cancer Immunology and Immunotherapy: Current Status and Future Perspectives." *Int Rev Cell Mol Biol* 331:1-53. doi: 10.1016/bs.ircmb.2016.09.008.

Deeks, E. D., and G. M. Keating. 2018. "Afatinib in advanced NSCLC: a profile of its use." *Drugs Ther Perspect* 34 (3):89-98. doi: 10.1007/s40267-018-0482-6.

Dhillon, S. 2019. "Neratinib in Early-Stage Breast Cancer: A Profile of Its Use in the EU." *Clin Drug Investig* 39 (2):221-229. doi: 10.1007/s40261-018-0741-2.

Dickler, M. N., S. M. Tolaney, H. S. Rugo, J. Cortes, V. Dieras, D. Patt, H. Wildiers, C. A. Hudis, J. O'Shaughnessy, E. Zamora, D. A. Yardley, M.

Frenzel, A. Koustenis, and J. Baselga. 2017. "MONARCH 1, A Phase II Study of Abemaciclib, a CDK4 and CDK6 Inhibitor, as a Single Agent, in Patients with Refractory HR(+)/HER2(-) Metastatic Breast Cancer." *Clin Cancer Res* 23 (17):5218-5224. doi: 10.1158/1078-0432. CCR-17-0754.

Dirix, L. Y., I. Takacs, G. Jerusalem, P. Nikolinakos, H. T. Arkenau, A. Forero-Torres, R. Boccia, M. E. Lippman, R. Somer, M. Smakal, L. A. Emens, B. Hrinczenko, W. Edenfield, J. Gurtler, A. von Heydebreck, H. J. Grote, K. Chin, and E. P. Hamilton. 2018. "Avelumab, an anti-PD-L1 antibody, in patients with locally advanced or metastatic breast cancer: a phase 1b JAVELIN Solid Tumor study." *Breast Cancer Res Treat* 167 (3):671-686. doi: 10.1007/s10549-017-4537-5.

Dotan, E., C. Aggarwal, and M. R. Smith. 2010. "Impact of Rituximab (Rituxan) on the Treatment of B-Cell Non-Hodgkin's Lymphoma." *P T* 35 (3):148-57.

Finn, R. S., M. Martin, H. S. Rugo, S. Jones, S. A. Im, K. Gelmon, N. Harbeck, O. N. Lipatov, J. M. Walshe, S. Moulder, E. Gauthier, D. R. Lu, S. Randolph, V. Dieras, and D. J. Slamon. 2016. "Palbociclib and Letrozole in Advanced Breast Cancer." *N Engl J Med* 375 (20):1925-1936. doi: 10.1056/NEJMoa1607303.

Gee, J. M., J. F. Robertson, I. O. Ellis, and R. I. Nicholson. 2001. "Phosphorylation of ERK1/2 mitogen-activated protein kinase is associated with poor response to anti-hormonal therapy and decreased patient survival in clinical breast cancer." *Int J Cancer* 95 (4):247-54.

Gerber, D. E. 2008. "Targeted therapies: a new generation of cancer treatments." *Am Fam Physician* 77 (3):311-9.

Goetz, M. P., M. Toi, M. Campone, J. Sohn, S. Paluch-Shimon, J. Huober, I. H. Park, O. Tredan, S. C. Chen, L. Manso, O. C. Freedman, G. Garnica Jaliffe, T. Forrester, M. Frenzel, S. Barriga, I. C. Smith, N. Bourayou, and A. Di Leo. 2017. "MONARCH 3: Abemaciclib As Initial Therapy for Advanced Breast Cancer." *J Clin Oncol* 35 (32):3638-3646. doi: 10.1200/JCO.2017.75.6155.

Goss, P. E., J. N. Ingle, K. I. Pritchard, N. J. Robert, H. Muss, J. Gralow, K. Gelmon, T. Whelan, K. Strasser-Weippl, S. Rubin, K. Sturtz, A. C.

Wolff, E. Winer, C. Hudis, A. Stopeck, J. T. Beck, J. S. Kaur, K. Whelan, D. Tu, and W. R. Parulekar. 2016. "Extending Aromatase-Inhibitor Adjuvant Therapy to 10 Years." *N Engl J Med* 375 (3):209-19. doi: 10.1056/NEJMoa1604700.

Hammond, M. E., D. F. Hayes, M. Dowsett, D. C. Allred, K. L. Hagerty, S. Badve, P. L. Fitzgibbons, G. Francis, N. S. Goldstein, M. Hayes, D. G. Hicks, S. Lester, R. Love, P. B. Mangu, L. McShane, K. Miller, C. K. Osborne, S. Paik, J. Perlmutter, A. Rhodes, H. Sasano, J. N. Schwartz, F. C. Sweep, S. Taube, E. E. Torlakovic, P. Valenstein, G. Viale, D. Visscher, T. Wheeler, R. B. Williams, J. L. Wittliff, and A. C. Wolff. 2010. "American Society of Clinical Oncology/College Of American Pathologists guideline recommendations for immunohistochemical testing of estrogen and progesterone receptors in breast cancer." *J Clin Oncol* 28 (16):2784-95. doi: 10.1200/JCO.2009.25.6529.

Han, H. S., V. Dieras, M. Robson, M. Palacova, P. K. Marcom, A. Jager, I. Bondarenko, D. Citrin, M. Campone, M. L. Telli, S. M. Domchek, M. Friedlander, B. Kaufman, J. E. Garber, Y. Shparyk, E. Chmielowska, E. H. Jakobsen, V. Kaklamani, W. Gradishar, C. K. Ratajczak, C. Nickner, Q. Qin, J. Qian, S. P. Shepherd, S. J. Isakoff, and S. Puhalla. 2018. "Veliparib with temozolomide or carboplatin/paclitaxel versus placebo with carboplatin/paclitaxel in patients with BRCA1/2 locally recurrent/metastatic breast cancer: randomized phase II study." *Ann Oncol* 29 (1):154-161. doi: 10.1093/annonc/mdx505.

Hanahan, D., and R. A. Weinberg. 2000. "The hallmarks of cancer." *Cell* 100 (1):57-70.

Hanahan, D., and R. A. Weinberg. 2011. "Hallmarks of cancer: the next generation." *Cell* 144 (5):646-74. doi: 10.1016/j.cell.2011.02.013.

Harbeck, N., C. S. Huang, S. Hurvitz, D. C. Yeh, Z. Shao, S. A. Im, K. H. Jung, K. Shen, J. Ro, J. Jassem, Q. Zhang, Y. H. Im, M. Wojtukiewicz, Q. Sun, S. C. Chen, R. G. Goeldner, M. Uttenreuther-Fischer, B. Xu, M. Piccart-Gebhart, and L. UX-Breast 1 study group. 2016. "Afatinib plus vinorelbine versus trastuzumab plus vinorelbine in patients with HER2-overexpressing metastatic breast cancer who had progressed on one previous trastuzumab treatment (LUX-Breast 1): an open-label,

randomised, phase 3 trial." *Lancet Oncol* 17 (3):357-66. doi: 10.1016/
S1470-2045(15)00540-9.

Harris, L. N., N. Ismaila, L. M. McShane, F. Andre, D. E. Collyar, A. M.
Gonzalez-Angulo, E. H. Hammond, N. M. Kuderer, M. C. Liu, R. G.
Mennel, C. Van Poznak, R. C. Bast, D. F. Hayes, and Oncology
American Society of Clinical. 2016. "Use of Biomarkers to Guide
Decisions on Adjuvant Systemic Therapy for Women With Early-Stage
Invasive Breast Cancer: American Society of Clinical Oncology
Clinical Practice Guideline." *J Clin Oncol* 34 (10):1134-50. doi:
10.1200/JCO.2015.65.2289.

Hartkopf, A. D., F. A. Taran, M. Wallwiener, C. B. Walter, B. Kramer, E.
M. Grischke, and S. Y. Brucker. 2016. "PD-1 and PD-L1 Immune
Checkpoint Blockade to Treat Breast Cancer." *Breast Care (Basel)* 11
(6):385-390. doi: 10.1159/000453569.

Hickinson, D. M., T. Klinowska, G. Speake, J. Vincent, C. Trigwell, J.
Anderton, S. Beck, G. Marshall, S. Davenport, R. Callis, E. Mills, K.
Grosios, P. Smith, B. Barlaam, R. W. Wilkinson, and D. Ogilvie. 2010.
"AZD8931, an equipotent, reversible inhibitor of signaling by epidermal
growth factor receptor, ERBB2 (HER2), and ERBB3: a unique agent for
simultaneous ERBB receptor blockade in cancer." *Clin Cancer Res* 16
(4):1159-69. doi: 10.1158/1078-0432.CCR-09-2353.

Hortobagyi, G. N. 1998. "Treatment of breast cancer." *N Engl J Med* 339
(14):974-84. doi: 10.1056/NEJM199810013391407.

Hortobagyi, G. N., S. M. Stemmer, H. A. Burris, Y. S. Yap, G. S. Sonke, S.
Paluch-Shimon, M. Campone, K. L. Blackwell, F. Andre, E. P. Winer,
W. Janni, S. Verma, P. Conte, C. L. Arteaga, D. A. Cameron, K.
Petrakova, L. L. Hart, C. Villanueva, A. Chan, E. Jakobsen, A. Nusch,
O. Burdaeva, E. M. Grischke, E. Alba, E. Wist, N. Marschner, A. M.
Favret, D. Yardley, T. Bachelot, L. M. Tseng, S. Blau, F. Xuan, F.
Souami, M. Miller, C. Germa, S. Hirawat, and J. O'Shaughnessy. 2016.
"Ribociclib as First-Line Therapy for HR-Positive, Advanced Breast
Cancer." *N Engl J Med* 375 (18):1738-1748. doi: 10.1056/
NEJMoa1609709.

Horwitz, K. B., and W. L. McGuire. 1975. "Specific progesterone receptors in human breast cancer." *Steroids* 25 (4):497-505.

Hu, X., W. Huang, and M. Fan. 2017. "Emerging therapies for breast cancer." *J Hematol Oncol* 10 (1):98. doi: 10.1186/s13045-017-0466-3.

Iacopetta, D., Y. Rechoum, and S. A. Fuqua. 2012. "The Role of Androgen Receptor in Breast Cancer." *Drug Discov Today Dis Mech* 9 (1-2):e19-e27. doi: 10.1016/j.ddmec.2012.11.003.

Jiang, Q., S. Zheng, and G. Wang. 2013. "Development of new estrogen receptor-targeting therapeutic agents for tamoxifen-resistant breast cancer." *Future Med Chem* 5 (9):1023-35. doi: 10.4155/fmc.13.63.

Jordan, V. C., and B. W. O'Malley. 2007. "Selective estrogen-receptor modulators and antihormonal resistance in breast cancer." *J Clin Oncol* 25 (36):5815-24. doi: 10.1200/JCO.2007.11.3886.

Joshi H, and Press MF. 2018. "Molecular oncology of breast cancer." In *The Breast*, edited by Copeland EM Bland KI, Klimberg VS, Gradishar WJ. Philadelphia, PA: Elsevier.

Kabos, P., and V. F. Borges. 2010. "Fulvestrant: a unique antiendocrine agent for estrogen-sensitive breast cancer." *Expert Opin Pharmacother* 11 (5):807-16. doi: 10.1517/14656561003641982.

Keegan, N. M., J. P. Gleeson, B. T. Hennessy, and P. G. Morris. 2018. "PI3K inhibition to overcome endocrine resistance in breast cancer." *Expert Opin Investig Drugs* 27 (1):1-15. doi: 10.1080/13543784.2018. 1417384.

Lee, A., and M. B. A. Djamgoz. 2018. "Triple negative breast cancer: Emerging therapeutic modalities and novel combination therapies." *Cancer Treat Rev* 62:110-122. doi: 10.1016/j.ctrv.2017.11.003.

Lee, J. J., K. Loh, and Y. S. Yap. 2015. "PI3K/Akt/mTOR inhibitors in breast cancer." *Cancer Biol Med* 12 (4):342-54. doi: 10.7497/j.issn. 2095-3941.2015.0089.

Lemon, H. M. 1970. "Abnormal estrogen metabolism and tissue estrogen receptor proteins in breast cancer." *Cancer* 25 (2):423-35.

Li, S. G., and L. Li. 2013. "Targeted therapy in HER2-positive breast cancer." *Biomed Rep* 1 (4):499-505. doi: 10.3892/br.2013.95.

Litton, J. K., H. S. Rugo, J. Ettl, S. A. Hurvitz, A. Goncalves, K. H. Lee, L. Fehrenbacher, R. Yerushalmi, L. A. Mina, M. Martin, H. Roche, Y. H. Im, R. G. W. Quek, D. Markova, I. C. Tudor, A. L. Hannah, W. Eiermann, and J. L. Blum. 2018. "Talazoparib in Patients with Advanced Breast Cancer and a Germline BRCA Mutation." *N Engl J Med* 379 (8):753-763. doi: 10.1056/NEJMoa1802905.

Loibl, S., and L. Gianni. 2017. "HER2-positive breast cancer." *Lancet* 389 (10087):2415-2429. doi: 10.1016/S0140-6736(16)32417-5.

LoRusso, P. M., J. Li, A. Burger, L. K. Heilbrun, E. A. Sausville, S. A. Boerner, D. Smith, M. J. Pilat, J. Zhang, S. M. Tolaney, J. M. Cleary, A. P. Chen, L. Rubinstein, J. L. Boerner, A. Bowditch, D. Cai, T. Bell, A. Wolanski, A. M. Marrero, Y. Zhang, J. Ji, K. Ferry-Galow, R. J. Kinders, R. E. Parchment, and G. I. Shapiro. 2016. "Phase I Safety, Pharmacokinetic, and Pharmacodynamic Study of the Poly(ADP-ribose) Polymerase (PARP) Inhibitor Veliparib (ABT-888) in Combination with Irinotecan in Patients with Advanced Solid Tumors." *Clin Cancer Res* 22 (13):3227-37. doi: 10.1158/1078-0432.CCR-15-0652.

Lumachi, F., D. A. Santeufemia, and S. M. Basso. 2015. "Current medical treatment of estrogen receptor-positive breast cancer." *World J Biol Chem* 6 (3):231-9. doi: 10.4331/wjbc.v6.i3.231.

Ma, C. X., T. Reinert, I. Chmielewska, and M. J. Ellis. 2015. "Mechanisms of aromatase inhibitor resistance." *Nat Rev Cancer* 15 (5):261-75. doi: 10.1038/nrc3920.

Maiello, M. R., A. D'Alessio, S. Bevilacqua, M. Gallo, N. Normanno, and A. De Luca. 2015. "EGFR and MEK Blockade in Triple Negative Breast Cancer Cells." *J Cell Biochem* 116 (12):2778-85. doi: 10.1002/jcb.25220.

Maira, S. M., S. Pecchi, A. Huang, M. Burger, M. Knapp, D. Sterker, C. Schnell, D. Guthy, T. Nagel, M. Wiesmann, S. Brachmann, C. Fritsch, M. Dorsch, P. Chene, K. Shoemaker, A. De Pover, D. Menezes, G. Martiny-Baron, D. Fabbro, C. J. Wilson, R. Schlegel, F. Hofmann, C. Garcia-Echeverria, W. R. Sellers, and C. F. Voliva. 2012. "Identification and characterization of NVP-BKM120, an orally available pan-class I

PI3-kinase inhibitor." *Mol Cancer Ther* 11 (2):317-28. doi: 10.1158/1535-7163.MCT-11-0474.

Manso, L., F. Moreno, R. Marquez, B. Castelo, A. Arcediano, M. Arroyo, A. I. Ballesteros, I. Calvo, M. J. Echarri, S. Enrech, A. Gomez, R. Gonzalez Del Val, E. Lopez-Miranda, M. Martin-Angulo, N. Martinez-Janez, C. Olier, and P. Zamora. 2015. "Use of bevacizumab as a first-line treatment for metastatic breast cancer." *Curr Oncol* 22 (2):e51-60. doi: 10.3747/co.22.2210.

McCormack, P. L., and S. J. Keam. 2008. "Bevacizumab: a review of its use in metastatic colorectal cancer." *Drugs* 68 (4):487-506. doi: 10.2165/00003495-200868040-00009.

Minuti, G., and L. Landi. 2015. "MET deregulation in breast cancer." *Ann Transl Med* 3 (13):181. doi: 10.3978/j.issn.2305-5839.2015.06.22.

Monnot, G. C., and P. Romero. 2018. "Rationale for immunological approaches to breast cancer therapy." *Breast* 37:187-195. doi: 10.1016/j.breast.2017.06.009.

Moverare-Skrtic, S., A. E. Borjesson, H. H. Farman, K. Sjogren, S. H. Windahl, M. K. Lagerquist, A. Andersson, A. Stubelius, H. Carlsten, J. A. Gustafsson, and C. Ohlsson. 2014. "The estrogen receptor antagonist ICI 182,780 can act both as an agonist and an inverse agonist when estrogen receptor alpha AF-2 is modified." *Proc Natl Acad Sci U S A* 111 (3):1180-5. doi: 10.1073/pnas.1322910111.

Murai, J., S. Y. Huang, A. Renaud, Y. Zhang, J. Ji, S. Takeda, J. Morris, B. Teicher, J. H. Doroshow, and Y. Pommier. 2014. "Stereospecific PARP trapping by BMN 673 and comparison with olaparib and rucaparib." *Mol Cancer Ther* 13 (2):433-43. doi: 10.1158/1535-7163.MCT-13-0803.

Nami, B., H. Maadi, and Z. Wang. 2018. "Mechanisms Underlying the Action and Synergism of Trastuzumab and Pertuzumab in Targeting HER2-Positive Breast Cancer." *Cancers (Basel)* 10 (10). doi: 10.3390/cancers10100342.

Nanda, R., L. Q. Chow, E. C. Dees, R. Berger, S. Gupta, R. Geva, L. Pusztai, K. Pathiraja, G. Aktan, J. D. Cheng, V. Karantza, and L. Buisseret. 2016. "Pembrolizumab in Patients With Advanced Triple-Negative Breast

Cancer: Phase Ib KEYNOTE-012 Study." *J Clin Oncol* 34 (21):2460-7. doi: 10.1200/JCO.2015.64.8931.

Nathan, M. R., and P. Schmid. 2017. "A Review of Fulvestrant in Breast Cancer." *Oncol Ther* 5 (1):17-29. doi: 10.1007/s40487-017-0046-2.

Ohaegbulam, K. C., A. Assal, E. Lazar-Molnar, Y. Yao, and X. Zang. 2015. "Human cancer immunotherapy with antibodies to the PD-1 and PD-L1 pathway." *Trends Mol Med* 21 (1):24-33. doi: 10.1016/j.molmed.2014. 10.009.

Perez-Garcia, J., E. Munoz-Couselo, J. Soberino, F. Racca, and J. Cortes. 2018. "Targeting FGFR pathway in breast cancer." *Breast* 37:126-133. doi: 10.1016/j.breast.2017.10.014.

Piccart-Gebhart, M. J., M. Procter, B. Leyland-Jones, A. Goldhirsch, M. Untch, I. Smith, L. Gianni, J. Baselga, R. Bell, C. Jackisch, D. Cameron, M. Dowsett, C. H. Barrios, G. Steger, C. S. Huang, M. Andersson, M. Inbar, M. Lichinitser, I. Lang, U. Nitz, H. Iwata, C. Thomssen, C. Lohrisch, T. M. Suter, J. Ruschoff, T. Suto, V. Greatorex, C. Ward, C. Straehle, E. McFadden, M. S. Dolci, R. D. Gelber, and Team Herceptin Adjuvant Trial Study. 2005. "Trastuzumab after adjuvant chemotherapy in HER2-positive breast cancer." *N Engl J Med* 353 (16):1659-72. doi: 10.1056/NEJMoa052306.

Pierson, M., and J. Swann. 1991. "Sensitization to noise-mediated induction of seizure susceptibility by MK-801 and phencyclidine." *Brain Res* 560 (1-2):229-36.

Puzanov, I., J. Sosman, A. Santoro, M. W. Saif, L. Goff, G. K. Dy, P. Zucali, J. A. Means-Powell, W. W. Ma, M. Simonelli, R. Martell, F. Chai, M. Lamar, R. E. Savage, B. Schwartz, and A. A. Adjei. 2015. "Phase 1 trial of tivantinib in combination with sorafenib in adult patients with advanced solid tumors." *Invest New Drugs* 33 (1):159-68. doi: 10.1007/ s10637-014-0167-5.

Quintas-Cardama, A., H. Kantarjian, J. Cortes, and S. Verstovsek. 2011. "Janus kinase inhibitors for the treatment of myeloproliferative neoplasias and beyond." *Nat Rev Drug Discov* 10 (2):127-40. doi: 10. 1038/nrd3264.

Quintela-Fandino, M., A. Urruticoechea, J. Guerra, M. Gil, A. Gonzalez-Martin, R. Marquez, E. Hernandez-Agudo, C. Rodriguez-Martin, M. Gil-Martin, R. Bratos, M. J. Escudero, S. Vlassak, F. Hilberg, and R. Colomer. 2014. "Phase I clinical trial of nintedanib plus paclitaxel in early HER-2-negative breast cancer (CNIO-BR-01-2010/GEICAM-2010-10 study)." *Br J Cancer* 111 (6):1060-4. doi: 10.1038/bjc.2014.397.

Ravelli, A., G. Roviello, D. Cretella, A. Cavazzoni, A. Biondi, M. R. Cappelletti, L. Zanotti, G. Ferrero, M. Ungari, F. Zanconati, A. Bottini, R. Alfieri, P. G. Petronini, and D. Generali. 2017. "Tumor-infiltrating lymphocytes and breast cancer: Beyond the prognostic and predictive utility." *Tumour Biol* 39 (4):1010428317695023. doi: 10.1177/1010428317695023.

Reguera-Nunez, E., P. Xu, A. Chow, S. Man, F. Hilberg, and R. S. Kerbel. 2019. "Therapeutic impact of Nintedanib with paclitaxel and/or a PD-L1 antibody in preclinical models of orthotopic primary or metastatic triple negative breast cancer." *J Exp Clin Cancer Res* 38 (1):16. doi: 10.1186/s13046-018-0999-5.

Ring, A., D. Wheatley, H. Hatcher, R. Laing, R. Plummer, M. Uttenreuther-Fischer, G. Temple, K. Pelling, and D. Schnell. 2015. "Phase I Study to Assess the Combination of Afatinib with Trastuzumab in Patients with Advanced or Metastatic HER2-Positive Breast Cancer." *Clin Cancer Res* 21 (12):2737-44. doi: 10.1158/1078-0432.CCR-14-1812.

Robson, M., S. A. Im, E. Senkus, B. Xu, S. M. Domchek, N. Masuda, S. Delaloge, W. Li, N. Tung, A. Armstrong, W. Wu, C. Goessl, S. Runswick, and P. Conte. 2017. "Olaparib for Metastatic Breast Cancer in Patients with a Germline BRCA Mutation." *N Engl J Med* 377 (6):523-533. doi: 10.1056/NEJMoa1706450.

Romond, E. H., E. A. Perez, J. Bryant, V. J. Suman, C. E. Geyer, Jr., N. E. Davidson, E. Tan-Chiu, S. Martino, S. Paik, P. A. Kaufman, S. M. Swain, T. M. Pisansky, L. Fehrenbacher, L. A. Kutteh, V. G. Vogel, D. W. Visscher, G. Yothers, R. B. Jenkins, A. M. Brown, S. R. Dakhil, E. P. Mamounas, W. L. Lingle, P. M. Klein, J. N. Ingle, and N. Wolmark. 2005. "Trastuzumab plus adjuvant chemotherapy for operable HER2-

positive breast cancer." *N Engl J Med* 353 (16):1673-84. doi: 10.1056/NEJMoa052122.

Russo, A. E., D. Priolo, G. Antonelli, M. Libra, J. A. McCubrey, and F. Ferrau. 2017. "Bevacizumab in the treatment of NSCLC: patient selection and perspectives." *Lung Cancer (Auckl)* 8:259-269. doi: 10.2147/LCTT.S110306.

Sacha, T. 2014. "Imatinib in chronic myeloid leukemia: an overview." *Mediterr J Hematol Infect Dis* 6 (1):e2014007. doi: 10.4084/MJHID.2014.007.

Santen, R. J., H. Brodie, E. R. Simpson, P. K. Siiteri, and A. Brodie. 2009. "History of aromatase: saga of an important biological mediator and therapeutic target." *Endocr Rev* 30 (4):343-75. doi: 10.1210/er.2008-0016.

Schoffski, P., S. Cresta, I. A. Mayer, H. Wildiers, S. Damian, S. Gendreau, I. Rooney, K. M. Morrissey, J. M. Spoerke, V. W. Ng, S. M. Singel, and E. Winer. 2018. "A phase Ib study of pictilisib (GDC-0941) in combination with paclitaxel, with and without bevacizumab or trastuzumab, and with letrozole in advanced breast cancer." *Breast Cancer Res* 20 (1):109. doi: 10.1186/s13058-018-1015-x.

Scott, L. J. 2017. "Niraparib: First Global Approval." *Drugs* 77 (9):1029-1034. doi: 10.1007/s40265-017-0752-y.

Segovia-Mendoza, M., M. E. Gonzalez-Gonzalez, D. Barrera, L. Diaz, and R. Garcia-Becerra. 2015. "Efficacy and mechanism of action of the tyrosine kinase inhibitors gefitinib, lapatinib and neratinib in the treatment of HER2-positive breast cancer: preclinical and clinical evidence." *Am J Cancer Res* 5 (9):2531-61.

Shen, Y., F. L. Rehman, Y. Feng, J. Boshuizen, I. Bajrami, R. Elliott, B. Wang, C. J. Lord, L. E. Post, and A. Ashworth. 2013. "BMN 673, a novel and highly potent PARP1/2 inhibitor for the treatment of human cancers with DNA repair deficiency." *Clin Cancer Res* 19 (18):5003-15. doi: 10.1158/1078-0432.CCR-13-1391.

Siow, Z. R., R. H. De Boer, G. J. Lindeman, and G. B. Mann. 2018. "Spotlight on the utility of the Oncotype DX((R)) breast cancer assay." *Int J Womens Health* 10:89-100. doi: 10.2147/IJWH.S124520.

Slamon, D. J., G. M. Clark, S. G. Wong, W. J. Levin, A. Ullrich, and W. L. McGuire. 1987. "Human breast cancer: correlation of relapse and survival with amplification of the HER-2/neu oncogene." *Science* 235 (4785):177-82.

Slamon, D. J., P. Neven, S. Chia, P. A. Fasching, M. De Laurentiis, S. A. Im, K. Petrakova, G. V. Bianchi, F. J. Esteva, M. Martin, A. Nusch, G. S. Sonke, L. De la Cruz-Merino, J. T. Beck, X. Pivot, G. Vidam, Y. Wang, K. Rodriguez Lorenc, M. Miller, T. Taran, and G. Jerusalem. 2018. "Phase III Randomized Study of Ribociclib and Fulvestrant in Hormone Receptor-Positive, Human Epidermal Growth Factor Receptor 2-Negative Advanced Breast Cancer: MONALEESA-3." *J Clin Oncol* 36 (24):2465-2472. doi: 10.1200/JCO.2018.78.9909.

Sledge, G. W., Jr., M. Toi, P. Neven, J. Sohn, K. Inoue, X. Pivot, O. Burdaeva, M. Okera, N. Masuda, P. A. Kaufman, H. Koh, E. M. Grischke, M. Frenzel, Y. Lin, S. Barriga, I. C. Smith, N. Bourayou, and A. Llombart-Cussac. 2017. "MONARCH 2: Abemaciclib in Combination With Fulvestrant in Women With HR+/HER2- Advanced Breast Cancer Who Had Progressed While Receiving Endocrine Therapy." *J Clin Oncol* 35 (25):2875-2884. doi: 10.1200/JCO.2017. 73.7585.

Smith, C. L., and B. W. O'Malley. 2004. "Coregulator function: a key to understanding tissue specificity of selective receptor modulators." *Endocr Rev* 25 (1):45-71. doi: 10.1210/er.2003-0023.

Solinas, C., L. Carbognin, P. De Silva, C. Criscitiello, and M. Lambertini. 2017. "Tumor-infiltrating lymphocytes in breast cancer according to tumor subtype: Current state of the art." *Breast* 35:142-150. doi: 10.1016/j.breast.2017.07.005.

Somlo, G., P. H. Frankel, B. K. Arun, C. X. Ma, A. A. Garcia, T. Cigler, L. V. Cream, H. A. Harvey, J. A. Sparano, R. Nanda, H. K. Chew, T. J. Moynihan, L. T. Vahdat, M. P. Goetz, J. H. Beumer, A. Hurria, J. Mortimer, R. Piekarz, S. Sand, J. Herzog, L. R. Van Tongeren, K. V. Ferry-Galow, A. P. Chen, C. Ruel, E. M. Newman, D. R. Gandara, and J. N. Weitzel. 2017. "Efficacy of the PARP Inhibitor Veliparib with Carboplatin or as a Single Agent in Patients with Germline BRCA1- or

BRCA2-Associated Metastatic Breast Cancer: California Cancer Consortium Trial NCT01149083." *Clin Cancer Res* 23 (15):4066-4076. doi: 10.1158/1078-0432.CCR-16-2714.

Specenier, P., and J. B. Vermorken. 2013. "Cetuximab: its unique place in head and neck cancer treatment." *Biologics* 7:77-90. doi: 10.2147/BTT. S43628.

Tabernero, J., R. Bahleda, R. Dienstmann, J. R. Infante, A. Mita, A. Italiano, E. Calvo, V. Moreno, B. Adamo, A. Gazzah, B. Zhong, S. J. Platero, J. W. Smit, K. Stuyckens, M. Chatterjee-Kishore, J. Rodon, V. Peddareddigari, F. R. Luo, and J. C. Soria. 2015. "Phase I Dose-Escalation Study of JNJ-42756493, an Oral Pan-Fibroblast Growth Factor Receptor Inhibitor, in Patients With Advanced Solid Tumors." *J Clin Oncol* 33 (30):3401-8. doi: 10.1200/JCO.2014.60.7341.

Tolaney, S. M., H. Nechushtan, I. G. Ron, P. Schoffski, A. Awada, C. A. Yasenchak, A. D. Laird, B. O'Keeffe, G. I. Shapiro, and E. P. Winer. 2016. "Cabozantinib for metastatic breast carcinoma: results of a phase II placebo-controlled randomized discontinuation study." *Breast Cancer Res Treat* 160 (2):305-312. doi: 10.1007/s10549-016-4001-y.

Tsang, J. Y., W. L. Au, K. Y. Lo, Y. B. Ni, T. Hlaing, J. Hu, S. K. Chan, K. F. Chan, S. Y. Cheung, and G. M. Tse. 2017. "PD-L1 expression and tumor infiltrating PD-1+ lymphocytes associated with outcome in HER2+ breast cancer patients." *Breast Cancer Res Treat* 162 (1):19-30. doi: 10.1007/s10549-016-4095-2.

Turner, N. C., J. Ro, F. Andre, S. Loi, S. Verma, H. Iwata, N. Harbeck, S. Loibl, C. Huang Bartlett, K. Zhang, C. Giorgetti, S. Randolph, M. Koehler, M. Cristofanilli, and Paloma Study Group. 2015. "Palbociclib in Hormone-Receptor-Positive Advanced Breast Cancer." *N Engl J Med* 373 (3):209-19. doi: 10.1056/NEJMoa1505270.

Turner, N. C., M. L. Telli, H. S. Rugo, A. Mailliez, J. Ettl, E. M. Grischke, L. A. Mina, J. Balmana, P. A. Fasching, S. A. Hurvitz, A. M. Wardley, C. Chappey, A. L. Hannah, M. E. Robson, and Abrazo Study Group. 2019. "A Phase II Study of Talazoparib after Platinum or Cytotoxic Nonplatinum Regimens in Patients with Advanced Breast Cancer and

Germline BRCA1/2 Mutations (ABRAZO)." *Clin Cancer Res* 25 (9):2717-2724. doi: 10.1158/1078-0432.CCR-18-1891.

Vieira, A. F., and F. Schmitt. 2018. "An Update on Breast Cancer Multigene Prognostic Tests-Emergent Clinical Biomarkers." *Front Med (Lausanne)* 5:248. doi: 10.3389/fmed.2018.00248.

Wahba, H. A., and H. A. El-Hadaad. 2015. "Current approaches in treatment of triple-negative breast cancer." *Cancer Biol Med* 12 (2):106-16. doi: 10.7497/j.issn.2095-3941.2015.0030.

Wakeling, A. E. 2000. "Similarities and distinctions in the mode of action of different classes of antioestrogens." *Endocr Relat Cancer* 7 (1):17-28.

Wang, D. Y., R. Fulthorpe, S. N. Liss, and E. A. Edwards. 2004. "Identification of estrogen-responsive genes by complementary deoxyribonucleic acid microarray and characterization of a novel early estrogen-induced gene: EEIG1." *Mol Endocrinol* 18 (2):402-11. doi: 10.1210/me.2003-0202.

Wilson, F. R., M. E. Coombes, Q. Wylie, M. Yurchenko, C. Brezden-Masley, B. Hutton, B. Skidmore, and C. Cameron. 2017. "Herceptin(R) (trastuzumab) in HER2-positive early breast cancer: protocol for a systematic review and cumulative network meta-analysis." *Syst Rev* 6 (1):196. doi: 10.1186/s13643-017-0588-2.

Wolff, A. C., M. E. Hammond, D. G. Hicks, M. Dowsett, L. M. McShane, K. H. Allison, D. C. Allred, J. M. Bartlett, M. Bilous, P. Fitzgibbons, W. Hanna, R. B. Jenkins, P. B. Mangu, S. Paik, E. A. Perez, M. F. Press, P. A. Spears, G. H. Vance, G. Viale, D. F. Hayes, Oncology American Society of Clinical, and Pathologists College of American. 2013. "Recommendations for human epidermal growth factor receptor 2 testing in breast cancer: American Society of Clinical Oncology/College of American Pathologists clinical practice guideline update." *J Clin Oncol* 31 (31):3997-4013. doi: 10.1200/JCO.2013.50.9984.

Wuerstlein, R., R. Kates, O. Gluz, E. M. Grischke, C. Schem, M. Thill, S. Hasmueller, A. Kohler, B. Otremba, F. Griesinger, C. Schindlbeck, A. Trojan, F. Otto, M. Knauer, R. Pusch, N. Harbeck, and Austria Switzerland Wsg-Prime investigators in Germany. 2019. "Strong impact of MammaPrint and BluePrint on treatment decisions in luminal early breast cancer: results of the WSG-PRIMe study." *Breast Cancer Res Treat.* doi: 10.1007/s10549-018-05075-x.

Xie, B., D. H. Wang, and S. J. Spechler. 2012. "Sorafenib for treatment of hepatocellular carcinoma: a systematic review." *Dig Dis Sci* 57 (5):1122-9. doi: 10.1007/s10620-012-2136-1.

Yakes, F. M., J. Chen, J. Tan, K. Yamaguchi, Y. Shi, P. Yu, F. Qian, F. Chu, F. Bentzien, B. Cancilla, J. Orf, A. You, A. D. Laird, S. Engst, L. Lee, J. Lesch, Y. C. Chou, and A. H. Joly. 2011. "Cabozantinib (XL184), a novel MET and VEGFR2 inhibitor, simultaneously suppresses metastasis, angiogenesis, and tumor growth." *Mol Cancer Ther* 10 (12):2298-308. doi: 10.1158/1535-7163.MCT-11-0264.

Zhang, H., Z. Bao, H. Liao, W. Li, Z. Chen, H. Shen, and S. Ying. 2017. "The efficacy and safety of tivantinib in the treatment of solid tumors: a systematic review and meta-analysis." *Oncotarget* 8 (68):113153-113162. doi: 10.18632/oncotarget.22615.

Zhao, Q., A. B. Parris, E. W. Howard, M. Zhao, Z. Ma, Z. Guo, Y. Xing, and X. Yang. 2017. "FGFR inhibitor, AZD4547, impedes the stemness of mammary epithelial cells in the premalignant tissues of MMTV-ErbB2 transgenic mice." *Sci Rep* 7 (1):11306. doi: 10.1038/s41598-017-11751-7.

Zimmer, A. S., M. Gillard, S. Lipkowitz, and J. M. Lee. 2018. "Update on PARP Inhibitors in Breast Cancer." *Curr Treat Options Oncol* 19 (5):21. doi: 10.1007/s11864-018-0540-2.

BIOGRAPHICAL SKETCH

Sara Charmsaz

Affiliation: Endocrine Oncology Research Group, Department of Surgery, Royal College of Surgeons in Ireland, Dublin 2, Dublin, Ireland.

Education: PhD

Business Address: Royal College of Surgeons in Ireland, 31A York Street Dublin, Ireland

Research and Professional Experience: Breast Cancer

Professional Appointments: Research Fellow

Publications from the Last 3 Years:
1. *Charmsaz, S., Hughes, E., Bane, F. T., Tibbitts, P., Mcilroy, M., Byrne, C., Cocchiglia, S., Mcbryan, J., Hennessy, B. T., Dwyer, R. M., Kerin, M. J., Hill, A. D. & Young, L. S. 2017b. S100beta as a serum marker in endocrine resistant breast cancer. *BMC Med, 15,* 79. Impact factor: 7.249.
2. *Charmsaz, S., Al-Ejeh, F., Yeadon, T. M., Miller, K. J., Smith, F. M., Stringer, B. W., Moore, A. S., Lee, F. T., Cooper, L. T., Stylianou, C., Yarranton, G. T., Woronicz, J., Scott, A. M., Lackmann, M. & Boyd, A. W. 2017a. EphA3 as a target for antibody immunotherapy in acute lymphoblastic leukemia. *Leukemia,* 31, 1779-1787. Impact factor: 12.104.
3. *Charmsaz, S., Beckett, K., Smith, F. M., Bruedigam, C., Moore, A. S., Al-Ejeh, F., Lane, S. W. & Boyd, A. W. 2015. EphA2 Is a Therapy Target in EphA2-Positive Leukemias but Is Not Essential for Normal Hematopoiesis or Leukemia. *PLoS One,* 10, e0130692. Impact factor: 2.806.
4. Browne, A. L., Charmsaz, S., Vareslija, D., Fagan, A., Cosgrove, N., Cocchiglia, S., Purcell, S., Ward, E., Bane, F., Hudson, L., Hill, A. D., Carroll, J. S., Redmond, A. M. & Young, L. S. 2018. Network analysis of SRC-1 reveals a novel transcription factor hub which regulates endocrine resistant breast cancer. *Oncogene.* Impact factor: 8.459.
5. Ward, E., Vareslija, D., Charmsaz, S., Fagan, A., Browne, A. L., Cosgrove, N. S., Cocchiglia, S., Purcell, S. P., Hudson, L., Das, S., O'connor, D., O'halloran, P. J., Sims, A. H., Hill, A. D. & Young, L. S. 2018. Epigenome-wide SRC-1 mediated gene silencing

repenses cellular differentiation in advanced breast cancer. *Clin Cancer Res.* Impact factor: 9.619.

6. Day, B. W., Stringer, B. W., Al-Ejeh, F., Ting, M. J., Wilson, J., Ensbey, K. S., Jamieson, P. R., Bruce, Z. C., Lim, Y. C., Offenhauser, C., Charmsaz, S., Cooper, L. T., Ellacott, J. K., Harding, A., Leveque, L., Inglis, P., Allan, S., Walker, D. G., Lackmann, M., Osborne, G., Khanna, K. K., Reynolds, B. A., Lickliter, J. D. & Boyd, A. W. 2013. EphA3 maintains tumorigenicity and is a therapeutic target in glioblastoma multiforme. *Cancer Cell,* 23, 238-48. Impact factor: 23.214.

7. Day, B. W., Stringer, B. W., Spanevello, M. D., Charmsaz, S., Jamieson, P. R., Ensbey, K. S., Carter, J. C., Cox, J. M., Ellis, V. J., Brown, C. L., Walker, D. G., Inglis, P. L., Allan, S., Reynolds, B. A., Lickliter, J. D. & Boyd, A. W. 2011. ELK4 neutralization sensitizes glioblastoma to apoptosis through downregulation of the anti-apoptotic protein Mcl-1. *Neuro Oncol,* 13, 1202-12. Impact factor: 7.371.

8. Stringer, B. W., Bunt, J., Day, B. W., Barry, G., Jamieson, P. R., Ensbey, K. S., Bruce, Z. C., Goasdoue, K., Vidal, H., Charmsaz, S., Smith, F. M., Cooper, L. T., Piper, M., Boyd, A. W. & Richards, L. J. 2016. Nuclear factor one B (NFIB) encodes a subtype-specific tumour suppressor in glioblastoma. *Oncotarget,* 7, 29306-20. Impact factor: 5.168.

9. Charmsaz, S. & Boyd, A. W. 2017. Eph receptors as oncotargets. *Oncotarget,* 8, 81727-81728. Impact factor: 5.168.

10. Charmsaz, S., Scott, A. M. & Boyd, A. W. 2017c. Targeted therapies in hematological malignancies using therapeutic monoclonal antibodies against Eph family receptors. *Exp Hematol,* 54, 31-39. Impact factor: 2.820.

In: Advances in Medicine and Biology   ISBN: 978-1-53616-181-6
Editor: Leon V. Berhardt   © 2019 Nova Science Publishers, Inc.

Chapter 4

IN SILICO APPROACH
ON H1N1 TREATMENT

Filia Stephanie and Usman Sumo Friend Tambunan[*]
Department of Chemistry, Universitas Indonesia, Depok, Indonesia

ABSTRACT

H1N1, also recognized as Swine Flu, was disseminated first in 1918 and attributed as a pandemic by the World Health Organization in 2009. H1N1 is a subtype of influenza A virus which belongs to the *orthomyxovirus* category. It encodes eleven proteins: envelope proteins (haemagglutinin and neuraminidase), matrix proteins (M1 and M2), Viral RNA Polymerases (PB2, PB1, PB1-F2, PA, and PB), and nonstructural proteins (NS1 and NS2). Neuraminidase and haemagglutinin are the proteins that make a difference on H1N1 strain with other strain of Influenza A. While all of the proteins have a different role, several antiviral compounds to inhibit H1N1 activity have been discovered through *in silico* drug discovery. Computational simulation and bioinformatics have been favoured for years as reliable approach to identify and design a potent drug candidate. Molecular docking and molecular dynamics simulation are useful tools to perform high-throughput screening from various databases

[*] Corresponding Author's E-mail: usman@ui.ac.id.

to search a novel candidate for H1N1 inhibitor. In this chapter, several strategies to combat H1N1 through *in silico* approach along with the discovered drug candidates are reviewed, such as neuraminidase inhibitors, M2 proton channel blocker, and RNA polymerase-complex inhibitor to interfere with H1N1 virus replication.

Keywords: influenza A, H1N1, *in silico*

INTRODUCTION

H1N1 belongs to the class of influenza virus A and commonly called swine flu. Swine influenza was first isolated from pigs in early 1920 and later confirmed to be able to infect human in 1974 when the strain was isolated from man lung tissue (Smith et al. 2010). H1N1 had never been classified as a virus that can cause infection in a human until the 2009 outbreak. The outbreak emerged in the Mexico City when the US Centers for Disease Control and Prevention (CDC) found a new case of respiratory disease in The Southern California caused by H1N1 (Ginsberg et al. 2009). The 2009 outbreak of H1N1 had caused 11,034 cases of infection in 41 countries before WHO determined the alert level to pandemic (Dotis and Roilides 2009). Infection of this virus is characterised by similar symptoms to other cases of influenzas, such as headache, sore throat, runny nose and high fever. Several cases of neurological disturbance and diarrhoea were also reported (Perez-Padilla et al. 2009; Cárdenas et al. 2014).

Several vaccines have been developed to protect humans since the 2009 outbreak, which classified into injection and nasal spray vaccine. Both of these types of vaccines contain the flu virus strain to force the human body to create antibodies (Girard et al. 2010). However, several adverse effects from muscle pain and nausea to bleeding disorders and Guillain-Barre Syndrome (GBS) have been reported upon the distribution of this vaccine subsequently (Vellozzi, Iqbal, and Broder 2014; He et al. 2010; Isai et al. 2012).

Influenza virus A, B, and C are classified into the *Orthomyxoviridae* family and can be identified by the segmented genomes with negative-stranded RNA. These viruses can be sphere or filament in shape (Shaw and Palese 2013). Influenza A virion contains hemagglutinin (HA) and neuraminidase (NA) glycoprotein in the surface and is classified into their subtypes based on the HA and NA. There are several influenza A subtypes that have been discovered, but only three HA (H subtypes 1, 2, and 3) and two NA (N subtypes 1, and 2) that are diagnosed to caused epidemics (Bouvier and Palese 2008). The influenza A envelope protein consists of HA, NA and matrix ion channel (M2) as the membrane proteins and M1 as the matrix protein. The nuclear export protein (NEP) and ribonucleoprotein complex (RNP) are located inside of the M1 matrix. RNP is composed of nucleoprotein (NP) and RNA-dependent RNA polymerase subunits called polymerase basic protein 1 (PB1), PB2 and polymerase acidic protein (PA) (Zebedee and Lamb 1988; Davis 2014).

Several approaches have been made to combat H1N1 infection and to increase immunity, such as vaccines and drugs that are available orally. Oral drugs to treat seasonal human influenza are classified into two classes: the NA inhibitor and the ion channel protein inhibitor. Oseltamivir and zanamivir are both NA inhibitor while adamantanes, which consists of rimantadine and amantadine, has a role in inhibiting the ion channel matrix protein (Dandagi and Byahatti 2011). Among the available drugs, oseltamivir is the most used drug to treat seasonal influenza. During the H1N1 outbreak in 2009, several cases of resistance to oseltamivir have been reported (Dharan et al. 2009; Hauge et al. 2009). Structural analysis has forecasted the H275Y mutation in NA can change the molecular conformation, making NA to be unrecognisable by the drug (Collins et al. 2009). Oseltamivir works by mimicking a sialic acid receptors and bind to the NA active site to obstruct the process of releasing the new viruses from the host cell. Beside the mutation, resistance to oseltamivir occurred because of the rearrangement of NA structure to hinder the attachment of oseltamivir sidechain (Moscona 2005).

To combat this resistance and develop new drug, Computer-aided Drug Design (CADD) is a useful tool for developing the structure of the novel drug candidate with suitable interaction to the target and supported by favourable pharmacological properties. This approach is divided into two general methods named structure-based drug design (SBDD) and ligand-based drug design (LBDD). While SBDD focuses on analysis of the macromolecular target, LBDD focuses on the known ligands to determine its critical pharmacological properties point. The primary purpose of these methods is to predict the structure-activity relationship between the ligand and the macromolecule target in order to design a new candidate for drug lead (Kapetanovic 2008; Jain 2017; Yu and Mackerell 2017). Molecular docking and dynamic simulation are mathematic-based algorithm tools that is commonly used in CADD methods. Molecular docking simulation is a method to predict a binding mode of a small molecule (ligand) to the macromolecule target (usually a protein, DNA or other macromolecules) by utilising conformational search and scoring function (Salmaso and Moro 2018). To predict the stability of the macromolecule-ligand complex, a more sophisticated technique like molecular dynamics simulation is often used in CADD to determine macromolecule-ligand interaction behaviour under specific conditions (Vlachakis et al. 2014).

CADD has been harnessed in the effort of H1N1 drug discovery. In this chapter, various research related to computational approach in the searching for novel inhibitor for H1N1 influenza A virus along with the discovered potential drug candidate will be discussed and summarised.

NEURAMINIDASE-TARGETED DRUG DISCOVERY

The NA belongs to the exosialidase category with EC number 3.2.1.18. NA works by separating N-acetylneuraminic acid and the neighbouring sugar residue through the cleavage of α-ketosidic linkage (Varghese and Colman 1991). Located in the surface of H1N1 virus along with HA and M2 protein, NA plays a central role in the development of new virus particles in the surface of the host cell to avoid aggregation. NA also helps in the movement of the virus to the host cell by cleaving neuraminic acid from the respiratory mucus (Shtyrya, Mochalova, and Bovin 2009).

NA consists of four indistinguishable polypeptides, each contains of approximately 470 amino acids. The NA binding site sequence is highly conserved and nine different NA subtypes are identified to have similar important amino acid residues in its active site (Arg118, Asp151, Arg152, Arg224, Glu276, Arg292, Arg371, and Tyr406) (Air 2012). Although NA is a tetramer with identical units, the monomeric structure does not have any cleavage activity (Saito, Taylor, and Webster 1995). This characteristic can be explained by looking at the propeller structure of NA in amino acids 353-403 that cannot be stabilized unless a disulfide bond is formed. Without the disulfide bond in the tetramer formation, the appropriate conformation cannot be constructed (Shtyrya, Mochalova, and Bovin 2009).

NA inhibitor is a type of medication that was used to treat seasonal influenza infection. NA inhibitor was created to mimic sialic acid in order to prevent a new virion from being released from the host cell (Mckimm-Breschkin 2013). Zanamivir and oseltamivir are the two most recognized substances in this class. Zanamivir and oseltamivir work by inhibiting the cellular release of the virus, preventing the replication of the virus (Mondal 2011). However, resistance cases for NA inhibitor have emerged. In the 2009 H1N1 outbreak, approximately 98.5% from total influenza A strains isolated from infected human were found to be resistant to oseltamivir in The United States. This number had increased rapidly compared to the 2007 statistic where only 0.7% of the strains were found resistant (Poland, Jacobson, and Ovsyannikova 2009). Resistance to zanamivir has also reported through several cases (Hurt et al. 2009; Gubareva et al. 2002). While arguments about the effectiveness of monotherapy usage in oseltamivir/zanamivir have developed, an effort to combine these two drugs were once made but resulted in low efficacy (Pizzorno et al. 2014). Several cases also showed the isolated strain of influenza virus that resistant to both oseltamivir and zanamivir (Escuret et al. 2008; Hurt and Barr 2008).

Laninamivir and Peramivir are two newly discovered NA inhibitors that have been approved to be used for influenza A treatment in Japan (Yamashita 2010; Shetty and Peek 2012). However, these two compounds are still undergone clinical trial by the US Food and Drug Administration. Laninamivir is administered by inhalation of its prodrug form, laninamivir octanoate. Meanwhile, peramivir is administered through an intravenous infusion (Leang et al. 2014). A study to compare the effectiveness of oseltamivir, zanamivir, laninamivir, and peramivir has been done in the 2010-2011 Japan influenza season to 45 patients with H1N1pdm09 infection as the treatment group. Peramivir was observed to be quicker in bringing down a fever to below 37.5°C compared to other medication (Shobugawa et al. 2012). Laninamivir also proved to be more effective than oseltamivir after being tested on influenza-infected mice (Tomozawa et al. 2019). Although this drug candidate shows high inhibition activity towards NA, several cross-resistance cases have been reported. Faster laninamivir dissociation was observed in its interaction to zanamivir-resistance NA (McKimm-Breschkin and Barrett 2015).

In order to overcome the resistance and develop novel NA inhibitor for H1N1 treatment, several *in silico* approaches have been done and the recent discovery will be summarized in this section.

Discovered NA Inhibitor from Oseltamivir/ Zanamivir Modification

Research on oseltamivir modification has been done by our lab to search for a potential NA inhibitor to overcome oseltamivir resistance. Bioinformatics method was conducted to retrieve a modified structure of oseltamivir with a various functional group. After conversion to the 3D structure, the newly designed compounds were docked into His274Tyr mutated NA. The 3D protein structure of this mutated NA was gained from sequence alignment and homology modelling. Three best ligands with good interaction with mutated NA were subjected to molecular dynamics simulation to evaluate the binding profile. The temperature of 312 K was chosen to mimic human temperature with fever. From the simulation, ligand AD3BF2D were identified to have a stable interaction with mutated NA and can be considered as an NA inhibitor candidate for in vitro/in vivo test before going through the clinical trial (Usman Sumo Friend Tambunan, Rachmania, and Parikesit 2015).

Zhi Xiang Zhao and team in 2019 researched on the discovery of novel NA inhibitor with SBDD. Pharmacophore features were profiled from oseltamivir and zanamivir derivatives, and the lead compound was generated from virtual screening of 670,000 compounds of the pharmacophore model database. Molecular docking and molecular dynamics simulation were subjected to 30 selected compounds with AMBER 12.0 to obtain the lead compound. From the simulation, lead compound 6a was obtained. Lead compound 6a was used as the template to design compound 6b to 6g. These compounds went through another molecular dynamics simulation to predict the interaction with NA. After the analysis by using in silico method, an in vitro study was then employed to study the inhibition activity of the newly synthesized compound to NA. Compound 6e (Acetic acid, 2-[[3-(4-morpholinyl) propyl] amino]-2-oxo-2-[(2-hydroxyphenyl) methylene] hydrazide) is identified to have the most inhibition potential with IC50 value of 2.37 µM (Z. X. Zhao et al. 2019).

Zhen Wang and team in 2017 designed several N-substituted oseltamivir derivatives using LBDD. Major features needed for NA inhibition were studied using 3D contour maps of Comparative Molecular Field Analysis (CoMFA) and Comparative Molecular Similarity Index Analysis (CoMSIA) from 65 oseltamivir derivative structures. The binding process of inhibitor and NA was identified to be impacted by hydrogen bond and hydrophobic interaction. By using the obtained information, a novel N-substituted oseltamivir was designed and evaluated with CADD methods such as 3D-QSAR, molecular docking, and molecular dynamics simulation. Designed compound with the best binding affinity to NA (compound 3) was synthesized and the inhibitory activity against NA was tested against oseltamivir as standard. Compound 3 was identified to have a better interaction with NA (Z. Wang et al. 2017). Li Ping Cheng and team in 2018 also made a similar approach to zanamivir structure to obtain a compound with a greater affinity towards NA. Two compounds, 45a and 45b, were chosen to be synthesized to assess the inhibition activity towards NA. From the result, compound 45b was observed to bind more effectively with NA compared to the zanamivir as reference (Cheng et al. 2018).

Kuanglei Wang and team in 2017 also evaluated NH_2-acyl oseltamivir analogues potential to be a novel NA inhibitor. Series of NH2-acyl oseltamivir derivatives were designed based on NA structure with 150 cavity and several NA inhibitor compounds with known inhibition activity. After the activity assay was conducted to synthesized NA inhibitor candidate, compound 11b were identified as a potent NA inhibitor. The CADD method of molecular docking simulation was employed to characterize the binding modes between compound 11b and NA. Compound 11b was confirmed to interact with NA with an additional hydrogen bond with Asp151 and two hydrogen bonds with Glu119 compared to oseltamivir. The new interactions indicated a more powerful binding pose with NA and higher NA inhibition potential. The newly-discovered compound also exhibited an excellent binding activity to NAs from H5N1 and H1N1, as well as the mutant (H1N1-H274Y and H5N1-H274Y) relative to the oseltamivir (K. Wang et al. 2017).

Discovered Novel NA Inhibitor

Different substance classes such as peptides and natural products can also be an alternative in the discovery of NA inhibitor. In 2012, our laboratory designed a cyclic peptide that can inhibit NA activity for H1N1 treatment. The amino acid sequences were retrieved with structure-based design. Analysis of the binding site and standard ligands (oseltamivir, zanamivir, and sialic acid as NA's natural substrate) interaction was conducted to observe the important amino acid residues involved in the binding with ligands. The 3D structure of H1N1 inhibitor was obtained through the homology modelling of the known sequence. In the designed inhibitor, the acetamin group has been known for its essential role in the interaction to the binding site as in sialic acid. Thus, asparagine (Asn/N) was used as one of the peptide monomers. From the oseltamivir's structure, etoxycarbonile was identified as one of the important group. Since ethoxycarbonyl is a modified form of the carboxyl group in the sialic acid (natural substrate of NA), aspartic acid was chosen. Tyrosine and lysine were also selected based on the interaction and the suitable size of their side chain.

The 'Entry Blocker' hypothesis was also used to add the possibility of the peptide sequence. The entry blocker peptide consists of six amino acids:

arginine-lysine-alanine-valine-leucine-proline. Eight peptides were acquired as the ligands, consisting of four ligands from the structure-based design and four ligands from the entry blocker amino acids. Three amino acids in the peptide were designed with probability and proximity in estimation. The sequence for the peptide from the structure-based design is DNY, NNY, DYY, and DDY, while ligands from the entry blocker are RRR, RRP, RPR, and LRL. Two cysteines were added into the sequence to obtain a cyclic peptide. The linear peptide can be cleaved with protease easily in the human body; thus nearly all peptide-based drugs is often designed as a cyclic peptide. The 3-dimension structure of the designed peptides was obtained with ACD labs.

The inhibition activity from the designed peptides were tested against NA using the molecular docking simulations to screen the ligands based on their poses and binding affinity to NA. Seven amino acid residues in NA catalytic site (Arg118, Asp151, Glu278, Arg292, Arg368, Tyr402, and Glu245) were used as the target site for the docking simulation. The ligands interaction with NA were then compared with standards and went through a drug-likeness screening with Lipinski's rule of five and Veber's Rules to determine their bioactivity. Visualization of the interaction was also created using PyMol, to get a better understanding of the NA catalytic site and the compatibility with ligands. Three designed peptide DNY, NNY, and LRL were identified as the best candidate for peptide NA inhibitor after the toxicological properties assessment and characterizing the Synthetic Accessibility (SA) (Usman Sumo Friend Tambunan 2012).

Our approach to discover a novel peptide NA inhibitor was continued. In the 2014 study, cyclic hexapeptides were designed as an NA inhibitor candidate. Each peptide contained two cysteines at the beginning and the end of the sequence to form a disulfide bond, achieving the cyclization. The peptide was chosen because of its high variability and accessibility. The NA 3-dimension structure was obtained through molecular modelling using SwissModel software by implementing the known sequence (Genbank code: ACT79135.1) (Figure 1). Sequence determination was designed with the information of NA catalytic site. Since the binding site located around Glu119, Arg152, Arg156, Trp179, Ser180, Ile223, Arg225, Glu228, Ser247,

Tyr274, Glu277, and Asn295 (polar and non-polar residues), the designed peptides were also set to be a combination of four amino acids with a polar and non-polar side chain. Through a combination of 11 polar amino acids and 9 non-polar amino acids, 5096 cyclic hexapeptide ligands were obtained and subjected to a molecular docking simulation against NA to predict the binding affinity. Fifteen best ligands with lowest ΔGbinding compared to standards went through a pharmacological screening using some software such as ToxTree, Osiris DataWarrior, and FAF-Drugs. Two peptides, CRMYPC and CRNFPC, were identified as the best candidate with good interaction and pharmacological properties.

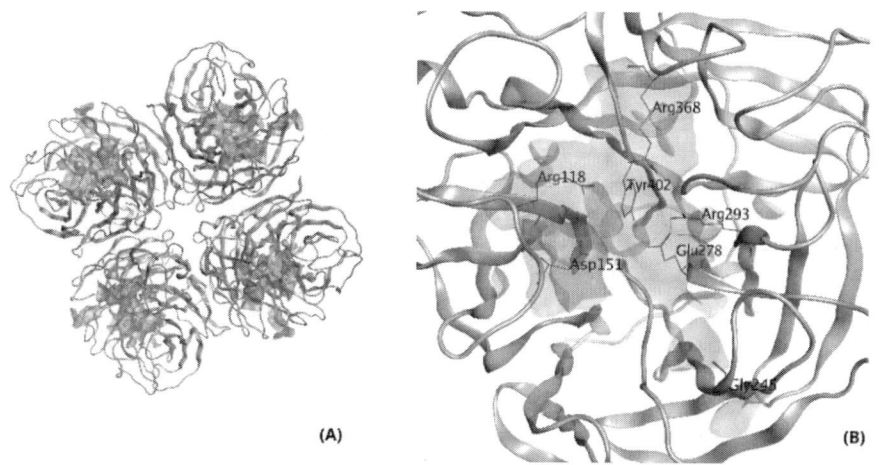

(A)   (B)

Figure 1. (A) Neuraminidase homotetramer and its binding site (magenta colour), modelled based on a known sequence from Genbank (ACT79135.1) using SwissModel (B) Prominent amino acid residues on the neuraminidase binding site used as the molecular docking target.

Molecular dynamics simulation was performed to characterize the stability of peptides-NA interaction. This simulation went under a variety of temperature (310 and 312 K) to observe a change of conformation in different temperature (human in normal and fever condition). The stability can be predicted with the time-dependent distance graph. From this study, the CRMYPC and NA interaction was observed to be stable since no conformational change was found. Hydrogen bond was formed with Arg368, Tyr402, and Glu425 in 310K, along with additional Asp151 in

312K. Thus, CRMYPC was selected as the best candidate for NA inhibitor in this study (Usman Sumo Friend Tambunan et al. 2014).

An effort to obtain a novel inhibitor for NA has also been made by Ganuskh Enkhtaivan and team, in 2018. In this study, the natural product was used as the inhibitor candidate. Berberine (an isoquinoline alkaloid) can be found on various plants and has been used in the traditional medicine for past decades. Berberine is proved to have a broad range of pharmacological features. The structure of berberine was used as a scaffold to do a structural modification in order to obtain a new NA inhibitor. Berberine-Piperazine conjugate derivatives were designed with the various active group, and the inhibition activity was investigated through *in silico* and in vitro analysis (Enkhtaivan et al. 2018).

M2 PROTON CHANNEL BLOCKER

The M2 proton channel is a homotetramer protein with 97 residues in each of its every monomer (Sakaguchi et al. 2002). This protein is a frequent target for influenza treatment because it plays a vital function in the virus life cycle. The M2 proton channel facilitates the release of viral RNA into the host cell, that is essential for virus replication (Pinto and Lamb 2006). The selectivity of this protein is very high and regulated by pH. There are two pH conditions during the virus entry into the host cell through the endosome, the first condition from extracellular pH to early endosomal pH, around 6; and the following condition during the late endosomal pH, around 5. The acidic condition activates the M2 channel to transfer protons across the envelope of the virus (Pielak and Chou 2011).

The M2 proton channel consists of four α-helices arranged in parallel to make up the channel pore. This protein is divided into three sections, extracellular N-terminal segment, transmembrane, and intracellular C-terminal segment (Schnell and Chou 2008). There are five essential residues in the M2 protein transmembrane domain, Val27, Ala30/Ser31, Gly34, His37, and Trp41. His37 plays a role as pH sensor, and Trp41 side chain acts as an essential gateway to proton transfer (Venkataraman, Lamb, and Pinto

2005). Because of their vital role, both of these residues are often recognized as the M2 functional core.

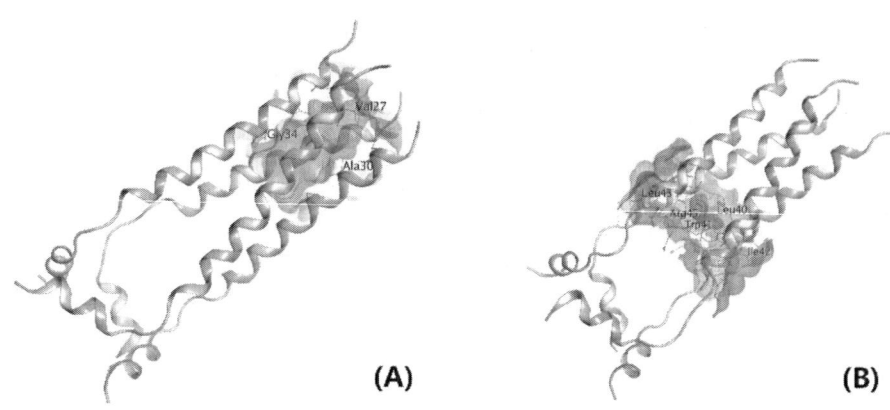

Figure 2. (A) Pore binding site of the H1N1 M2 proton channel (B) Surface binding site of the H1N1 M2 proton channel.

Several discovered compounds are known for its ability to interact with the M2 protein to block the proton channel. Adamantine (1-aminoadamantane hydrochloride) and Rimantadine (amantadine analogue, α-methyl-1-adamantane methylamine hydrochloride) are two anti-influenza drugs which have been widely used as proton channel blocker (Chizhmakov et al. 1996). For years, the interaction between these drugs and M2 protein was suggested to happen in the M2 pore binding site, which consists of Val27, Ala30/Ser31, and Gly34 (Acharya et al. 2010). After surface Nuclear Magnetic Resonance (sNMR) structure analysis of M2-rimantadine complex discovered in 2008, another binding site was revealed. Rimantadine was found to bind to the C-terminal of the transmembrane domain surface binding site constituted of Leu40, Leu43, Asp44, Ile42, and Arg45 (Schnell and Chou 2008). The visualization of both binding sites of the M2 proton channel is presented in Figure 2. Several computational approaches have been done in order to study the energetics of pore and surface binding sites. Through molecular dynamics simulation, the interaction between the adamantane-based drugs and M2 protein in the pore binding site was observed to be thermodynamically stable in the physiological pH compared to the surface binding site (Gu et al. 2011). Another study using computational solvent mapping also revealed a weaker hot spots in the interaction of rimantadine and surface binding site. Therefore, the pore binding site (the internal cavity) was identified as the primary pharmacological hot spots for rimantadine (Chuang et al. 2009).

The efficacy of adamantine and rimantadine-based drugs has been proved to be declined as several cases of resistance were reported. Substitution of serine 31 with asparagine, S31N, is the most common resistance in influenza A virus. These substitutions were found in H1N1, H3N2, and H1N2 (Krumbholz et al. 2009). Stelios Eleftheratos and team in 2010 did an interaction study of the aminoadamantane derivatives and M2 proton channel in its pore binding site. Several derivatives of aminoadamantanes were subjected through a series of molecular docking simulation to analyze the binding mode with M2 proton channel. From the result, Val27, Ala30, and Ser31 were observed to be the essential residues in M2 pore binding site. Ser31 were identified to be the most stable and were able to connect with the amino group in the inhibitor through hydrogen bonding (Eleftheratos et al. 2010). With the replacement of Ser31 to Asn31, adamantine-based drugs are no longer a promising drug for influenza A. In addition to S31N; the L26F and V27A substitution in influenza A M2 proton channel protein were also reported (Gordon et al. 2017). In order to combat drug resistance, several *in silico* approach has also been done.

Discovery of Amantadine-Based M2 Proton Channel Inhibitor

Our laboratory has made several efforts in searching for a drug candidate for M2 proton channel inhibitor. In 2013, we designed 1,447 amantadine-based compounds for virtual screening against the M2 proton channel inhibitor. Structure-activity relationship of amantadine was used to substitute the amino groups of amantadine with bicyclic and tricyclic amines to obtain the compound library. These compounds were then subjected to a molecular docking simulation to observe the interaction to the M2 proton channel. The 3D structure of H1N1 M2 protein was retrieved from homology modelling, using M2 channel 2RLF as the structure template.

Multiple sequence alignment was performed to obtain a representative sequence of H1N1 M2 channel for the main protein target in molecular docking simulation.

From the simulation, three amantadine-based compounds were identified to have a better inhibition activity to the M2 proton channel compared to adamantine and rimantadine as the standards. The three selected compounds are [hydroxyl (1H-imidazole-4-yl) methyl] tricyclo[3.3.1..03,7]non-3-ylcarbamic acid, (tricyclo[3.3.1.13,7]dec-1-ylimino) bis (1H-imidazole-4-ylmethanol), and (tricyclo [3.3.1.03,7]non-3-ylimino) bis (1H-imidazole-4-ylmethanol). These compounds were known to bind with the functional residue Asp44 through hydrogen bonds, located in the protein exit of the M2 channel. The binding of ligands to this residue will not only place the channel into a closed conformation but also block the proton exits. Meanwhile, amantadine and rimantadine were observed unable to form hydrogen bonds with this residue.

The molecular dynamics simulation was executed using MOE 2008.10 using MMFF94x as the force field and Born solvation model to assess the stability of hydrogen bonds between the ligands and M2 proton channel. This simulation was conducted under two temperatures, 300K as the default temperature and 312K to mimic fever condition. As a result, the hydrogen bonds were still formed between the three ligands and Asp44 residue of the M2 proton channel. Therefore, these three compounds can be used as a potential inhibitor which needs to go through further assessment (Tambunan 2013).

Yibing Wu and team in 2014 have done other effort in the M2 proton channel inhibitor discovery. After the discovery of the potential benzyl-substituted amantadines on inhibiting influenza A M2 proton channel with 240-fold efficiency compared to amantadine (X. Zhao et al. 2011), a study on searching a dual-inhibitor candidate for M2 channel wild type (WT) and S31N has been done. Modification of aryl amantadine with the conjugation of –CH2-heteroaryl to the amino group was done to acquire the inhibitory activity against S31N mutant. As a result of the modification, several isoxazone-adamantane and oxadiazole derivatives were designed and went through structure-activity analysis. After the inhibitory activity was confirmed, molecular dynamics simulation were conducted to analyze the interaction between the protein and ligands. Hydrophobic interaction was observed with Val27 and Asn31 side chain residues. From this study, compound M2WJ322 was identified to have a stable hydrogen bonding to Asn31 side chain in a preferable position (J. Wang et al. 2013).

Discovery of Novel M2 Proton Channel Inhibitor

Our laboratory has done modification of a discovered inhibitor of M2 proton channel to obtain a compound with a better affinity towards the protein in 2012. (1R, 2R, 3R, 5S)-(-)-isopinocampheylamine was first identified to be able to block the M2 protein channel (X. Zhao et al. 2011). The newly discovered compound and amantadine have an amino group that plays a key role to be a barrier of proton flow in influenza A virus M2 proton channel. In order to modify this compound, several active groups from a previously synthesized inhibitor of M2 channel were chosen as a guideline. Fifty-two ligands were obtained from this modification with the ACDLabs software. After the energy minimization and optimization was done, these compounds were subjected to a molecular docking simulation to screen the potential candidate with suitable binding modes to M2 proton channel. In this process, three ligands were used as the standard compounds (amantadine, rimantadine, and isopinocampheylamine). A rigid docking protocol was applied to the protein, while the ligands were made flexible to pinpoint the optimum binding pose. From this simulation, compound B18, B20, and A20 were identified to have a better binding mode with the M2 proton channel compared to standards. The ΔGbinding and pKi (inhibition constant) were two parameters used in comparing the result. Drug scan analysis gave an exceptional drug-likeness properties for all three chosen ligands, and pharmacological assessment confirmed these ligands are not carcinogenic and mutagenic (U. S.F. Tambunan, Harganingtyas, and Parikesit 2012).

Acylguanidines derivatives have also been studied for its ability to inhibit the M2 proton channel. Acylguanidines act as an inhibitor for viroporins of several viruses, including hepatitis virus and HIV-1(Ewart et al. 2002; Premkumar et al. 2004)· Viroporins are small and hydrophobic transmembrane proteins that facilitate the movement of small molecules and are essential for the viral replication (Nieva, Madan, and Carrasco 2012). The M2 proton channel is the viroporin of influenza A virus. In 2016, Pouria H. Jalily and team discovered a potent inhibitor for influenza A M2 viroporin that can inhibit both the WT and mutant M2 protein from the acylguanidines derivatives. Four compounds of acylguanidines were subjected to cell assay to characterize their inhibition activity, and from the result, hexamethylene amiloride was observed to have the best activity among other acylguanidines. The important functional groups of hexamethylene amiloride were analyzed to give insight about the structure-activity relationship and substitution was implemented to enhance the inhibition activity to obtain a novel compound with dual-inhibition ability. Molecular docking simulation was performed to both WT and S31N mutant of the M2 proton channel to study the interaction between ligands and the target protein. From the result, compounds 26 and 27 were obtained as an inhibitor to both WT and S31N mutant M2 proton channel (Jalily et al. 2016).

Drug repurposing has also been proposed in the discovery of the M2 proton channel. Drug repurposing is a strategy to find a new purpose/indication for the already-available drugs. This process is often conducted as a shortcut to the common path on the traditional drug discovery (Pushpakom et al. 2018). There are several approaches in a drug repurposing, including computational approach, biological approach, or even a combination of both approaches (Xue et al. 2018)· Draginja Radosevic and team in 2019 has made an effort of drug repurposing through a computational approach. About 2,627 small compounds from DrugBank that have been approved to be used as a drug underwent virtual screening as the influenza A M2 proton channel inhibitor. Virtual screening library was build using ChEMBL Target Report Card to remove duplicates and inactive compounds. The EIIP/AQVN value and ligand-based virtual screening were also employed to filter the M2 inhibitor candidates. A similar value of EIIP/AQVN describes the similarity of their therapeutic properties

(Veljkovic et al. 2013). Thirty-nine selected compounds underwent ligand-based virtual screening and resulted in 5 best candidates.

In order to obtain a drug structure that can inhibit both WT and S31N mutant, the five chosen candidates were subjected to a molecular docking simulation. From the result, cycrimine was observed to have the lowest binding energy with the M2 proton channel protein. This compound also exhibited a good interaction with Ser31 (in WT) and Asn31 (in S31N) mutant by forming a hydrogen bond. Cycrimine was also identified to have a hydrophobic interaction with Ala30. Validation of this study was achieved through an inhibition activity assay, using merimepodib against influenza A as the positive control (Radosevic et al. 2019).

RNA-POLYMERASE COMPLEX INHIBITOR

PA, PB1, and PB2 are the building blocks of the RNA polymerase complex in the influenza virus (Vasin et al. 2014). It is one of the highly conserved parts in the virus that contains various potential for anti-influenza development (Babar, Zaidi, and Tahir 2014). Several approaches in targeting the RNA polymerase complex includes RNA synthesis inhibitor, cap-snatching inhibitor, and the subunits polymerase interaction-targeted inhibitor (Shen, Lou, and Wang 2015). All RNA polymerase inhibitors have an impact on the disruption of the virus transcription and replication.

There are several compounds with known activity towards the RNA polymerase complex of influenza A virus, such as Ribavirin, Favipiravir, and Baloxavir Acid. Ribavirin is one of the first RNA synthesis inhibitors. It was approved on the US as a broad spectrum drug which mainly targeting hepatitis C virus and the human respiratory syncytial virus (Rowe et al. 2010). As the pandemic of H1N1 emerged in 2009, study about the intravenous use of ribavirin was conducted to analyze the efficacy and resistance-potential of the drug (Chan-Tack, Murray, and Birnkrant 2009). Favipiravir (6-fluoro-3-hydroxy-2-pyrazinecarboxamide) is the recently discovered drug that can act as a substrate to the RNA polymerase complex, resulted in inhibition of the protein. The efficacy of favipiravir has been studied for the 2009 pandemic of the H1N1, H5N1, and H7N9 strains (Furuta et al. 2013). Although favipiravir still undergoes clinical trial, a recent study on potential K229R mutation generated from the drug use has been observed (Goldhill et al. 2018). Baloxavir acid is another drug candidate for influenza that is known to inhibit the cap-dependent endonuclease in the RNA polymerase complex (Yang 2019). This drug was tested as its prodrug form, baloxavir marboxil. Baloxavir was shown to have a broad range activity towards the various strain of influenza A. The safety and efficacy of this drug were still under the investigation on a clinical trial (Noshi et al. 2018).

While there is no approved drug to inhibit the RNA polymerase of H1N1 specifically, there are several studies in the discovery of this protein complex inhibitor using a computational approach. Arundhathi Arivajiagane recently developed a novel peptide inhibitor targeting the PAn-PB1c interaction in the RNA polymerase complex. The protein-protein interaction was first investigated through the hotspot prediction. The peptide inhibitor was selected using the peptide library, and 122 peptides were generated for further study. All the selected peptides underwent a docking simulation with FlexPepDock against the 3D structure of the RNA polymerase complex to obtain peptide with the best binding affinity and disturbance on the protein-protein interaction. From the study, FluAPep1 (Ser-Arg-Ala-Arg-Ile-Asp-Ala-Arg-Ile) were observed to have a good affinity towards the RNA polymerase complex and underwent further modification to enhance the inhibition activity (Arivajiagane et al. 2019).

Another inhibitor targeting protein-protein interaction has also discovered by Cristina Tintori and team in 2014. High throughput molecular docking simulation was done through the virtual screening of potential inhibitor that can disturb the PA-PB1 protein interaction. First, the molecular dynamics simulation was employed to the 3D structure of the protein complex to identify the important residues in those proteins interaction. After that, 703,200 compounds from Asinex database were screened using a high throughput molecular docking simulation to search molecules that can mimic the interaction by the first part of PB1. From this simulation, several compounds with 3-cyano-4,6-diphenyl-pyridine nucleus were identified as a novel influenza A polymerase inhibitor (Tintori et al. 2014).

Approach on database screening was also made to search for a novel inhibitor against the influenza A RNA polymerase complex. Giulia Muratore and team in 2012 screened three million compounds from the ZINC database to obtain a molecule with similar fingerprint with C-terminal fragment of PA that was bound to PB1. From the virtual screening, 32 molecules were obtained and subjected to ELISA method (Muratore et al. 2012). Screening from the GVK Bioscience's Drug Database was done by Mayuko Fukuoka and team in 2012 using molecular docking simulation against Influenza A RNA polymerases, to target the PA-PB1 interaction. Benzbromarone, diclazuril, and trenbolone were obtained from the screening and underwent cell viability and plaque assay (Fukuoka et al. 2012).

CONCLUSION

Computational methods have played an essential role in the drug discovery; whether it is used for analyzing the interaction between drug and the target protein, or virtual screening of a drug candidate. Molecular docking and dynamic simulation are two of bioinformatics tool that is often used in this approach.

Several approaches to search a small molecule that can inhibit protein in H1N1 virus have been conducted through various *in silico* methods, including the discovery of compounds that inhibiting neuraminidase, targeting the M2 proton channel, and disturbing the RNA synthesis of the virus. These efforts resulted in some promising drug candidates for H1N1 treatment, although it needs to be further validated through the in vitro, in vivo, as well as the clinical trial study.

REFERENCES

Acharya, R., V. Carnevale, G. Fiorin, B. G. Levine, A. L. Polishchuk, V. Balannik, I. Samish, et al. 2010. "Structure and Mechanism of Proton Transport through the Transmembrane Tetrameric M2 Protein Bundle of the Influenza A Virus." *Proceedings of the National Academy of Sciences.* doi:10.1073/pnas.1007071107.

Air, Gillian M. 2012. "Influenza Neuraminidase." *Influenza and Other Respiratory Viruses.* doi:10.1111/j.1750-2659.2011.00304.x.

Arivajiagane, Arundhathi, Narendrakumar Ravi Varadharajulu, Kumar Seerangan, and Rajesh Rattinam. 2019. "In Silico Structure-Based Design of Enhanced Peptide Inhibitors Targeting RNA Polymerase PA N -PB1 C Interaction." *Computational Biology and Chemistry.* doi:10.1016/j.compbiolchem.2018.12.009.

Babar, Mustafeez Mujtaba, Najam Us Sahar Sadaf Zaidi, and Muhammad Tahir. 2014. "Global Geno-Proteomic Analysis Reveals Cross-Continental Sequence Conservation and Druggable Sites among Influenza Virus Polymerases." *Antiviral Research.* doi:10.1016/j.antiviral.2014.10.013.

Bouvier, Nicole M., and Peter Palese. 2008. "The Biology of Influenza Viruses." *Vaccine.* doi:10.1016/j.vaccine.2008.07.039.

Cárdenas, Graciela, José Luis Soto-Hernández, Alexandra Díaz-Alba, Yair Ugalde, Jorge Mérida-Puga, Marcos Rosetti, and Edda Sciutto. 2014. "Neurological Events Related to Influenza A (H1N1) Pdm09." *Influenza and Other Respiratory Viruses.* doi:10.1111/irv.12241.

Chan-Tack, Kirk M., Jeffrey S. Murray, and Debra B. Birnkrant. 2009. "Use of Ribavirin to Treat Influenza." *New England Journal of Medicine.* doi:10.1056/nejmc0905290.

Cheng, Li Ping, Tian Chi Wang, Rao Yu, Meng Li, and Jin Wen Huang. 2018. "Design, Synthesis and Biological Evaluation of Novel Zanamivir Derivatives as Potent Neuraminidase Inhibitors." *Bioorganic and Medicinal Chemistry Letters* 28 (23–24). Elsevier: 3622–29. doi:10.1016/j.bmcl.2018.10.040.

Chizhmakov, I. V., F. M. Geraghty, D. C. Ogden, A. Hayhurst, M. Antoniou, and A. J. Hay. 1996. "Selective Proton Permeability and PH Regulation of the Influenza Virus M2 Channel Expressed in Mouse Erythroleukaemia Cells." *Journal of Physiology*. doi:10.1113/jphysiol.1996.sp021495.

Chuang, Gwo Yu, Dima Kozakov, Ryan Brenke, Dmitri Beglov, Frank Guarnieri, and Sandor Vajda. 2009. "Binding Hot Spots and Amantadine Orientation in the Influenza A Virus M2 Proton Channel." *Biophysical Journal*. doi:10.1016/j.bpj.2009.09.004.

Collins, P. J., L. F. Haire, Y. P. Lin, J. Liu, R. J. Russell, P. A. Walker, S. R. Martin, et al. 2009. "Structural Basis for Oseltamivir Resistance of Influenza Viruses." *Vaccine*. doi:10.1016/j.vaccine.2009.07.017.

Dandagi, GirishL, and SujataM Byahatti. 2011. "An Insight into the Swine-Influenza A (H1N1) Virus Infection in Humans." *Lung India*. doi:10.4103/0970-2113.76299.

Davis, L. E. 2014. "Influenza Virus." In *Encyclopedia of the Neurological Sciences*. doi:10.1016/B978-0-12-385157-4.00381-X.

Dharan, Nila J., Larisa V. Gubareva, John J. Meyer, Margaret, Okomo-Adhiambo, Reginald C. McClinton, Steven A. Marshall, et al. 2009. "Infections with Oseltamivir-Resistant Influenza A(H1N1) Virus in the United States." *JAMA - Journal of the American Medical Association*. doi:10.1001/jama.2009.294.

Dotis, J., and E. Roilides. 2009. "H1N1 Influenza A Infection." *Hippokratia*.

Eleftheratos, Stelios, Philip Spearpoint, Gabriella Ortore, Antonios Kolocouris, Adriano Martinelli, Stephen Martin, and Alan Hay. 2010. "Interaction of Aminoadamantane Derivatives with the Influenza A Virus M2 Channel-Docking Using a Pore Blocking Model." *Bioorganic and Medicinal Chemistry Letters*. doi:10.1016/j.bmcl.2010.05.049.

Enkhtaivan, Ganuskh, Doo Hwan Kim, Gyun Seok Park, Muthuraman Pandurangan, Daniel A. Nicholas, So Hyun Moon, Avinash A. Kadam, Rahul V. Patel, Han Seung Shin, and Bhupendra M. Mistry. 2018. "Berberine-Piperazine Conjugates as Potent Influenza Neuraminidase Blocker." *International Journal of Biological Macromolecules*. doi:10.1016/j.ijbiomac.2018.08.047.

Escuret, Vanessa, Emilie Frobert, Maude Bouscambert-Duchamp, Murielle Sabatier, Isidore Grog, Martine Valette, Bruno Lina, Florence Morfin, and Olivier Ferraris. 2008. "Detection of Human Influenza A (H1N1) and B Strains with Reduced Sensitivity to Neuraminidase Inhibitors." *Journal of Clinical Virology*. doi:10.1016/j.jcv.2007.10.019.

Ewart, Gary D., Kerry Mills, Graeme B. Cox, and Peter W. Gage. 2002. "Amiloride Derivatives Block Ion Channel Activity and Enhancement of Virus-like Particle Budding Caused by HIV-1 Protein Vpu." In *European Biophysics Journal*. doi:10.1007/s002490100177.

Fukuoka, Mayuko, Moeko Minakuchi, Atsushi Kawaguchi, Kyosuke Nagata, Yuji O. Kamatari, and Kazuo Kuwata. 2012. "Structure-Based Discovery of Anti-Influenza Virus A Compounds among Medicines." *Biochimica et Biophysica Acta - General Subjects*. doi:10.1016/j.bbagen.2011.11.003.

Furuta, Yousuke, Brian B. Gowen, Kazumi Takahashi, Kimiyasu Shiraki, Donald F. Smee, and Dale L. Barnard. 2013. "Favipiravir (T-705), a Novel Viral RNA Polymerase Inhibitor." *Antiviral Research*. doi:10.1016/j.antiviral.2013.09.015.

Ginsberg, M., J. Hopkins, A. Maroufi, G. Dunne, D. R. Sunega, J. Giessick, P. McVay, et al. 2009. "Swine Influenza A (H1N1) Infection in Two Children – Southern California, March–April." *Morbidity and Mortality Weekly Report*.

Girard, Marc P., John S. Tam, Olga M. Assossou, and Marie Paule Kieny. 2010. "The 2009 A (H1N1) Influenza Virus Pandemic: A Review." *Vaccine*. doi:10.1016/j.vaccine.2010.05.031.

Goldhill, Daniel H., Aartjan J. W. te Velthuis, Robert A. Fletcher, Pinky Langat, Maria Zambon, Angie Lackenby, and Wendy S. Barclay. 2018. "The Mechanism of Resistance to Favipiravir in Influenza." *Proceedings of the National Academy of Sciences*. doi:10.1073/pnas.1811345115.

Gordon, Nathan A., Kelly L. McGuire, Spencer K. Wallentine, Gregory A. Mohl, Jonathan D. Lynch, Roger G. Harrison, and David D. Busath. 2017. "Divalent Copper Complexes as Influenza A M2 Inhibitors." *Antiviral Research*. doi:10.1016/j.antiviral.2017.10.009.

Gu, Ruo Xu, Limin Angela Liu, Dong Qing Wei, Jian Guo Du, Lei Liu, and Hong Liu. 2011. "Free Energy Calculations on the Two Drug Binding Sites in the M2 Proton Channel." *Journal of the American Chemical Society*. doi:10.1021/ja1114198.

Gubareva, Larisa V., Mikhail N. Matrosovich, Malcolm K. Brenner, Richard C. Bethell, and Robert G. Webster. 2002. "Evidence for Zanamivir Resistance in an Immunocompromised Child Infected with Influenza B Virus." *The Journal of Infectious Diseases*. doi:10.1086/314440.

Hauge, Siri H., Susanne Dudman, Katrine Borgen, Angie Lackenby, and Olav Hungnes. 2009. "Oseltamivir-Resistant Influenza Viruses A (H1N1), Norway, 2007-08." *Emerging Infectious Diseases*. doi:10.3201/eid1502.081031.

He, Xiao Li, Shu Qiong Kang, Cheng Ying Gong, and Guan Yao Jiang. 2010. "Safety Observation of Influenza a H1N1 Influenza Vaccine Vaccinations in 3300 Medical Workers." *Chinese Journal of Evidence-Based Medicine*.

Hurt, A. C., J. K. Holien, M. Parker, A. Kelso, and I. G. Barr. 2009. "Zanamivir-Resistant Influenza Viruses with a Novel Neuraminidase Mutation." *Journal of Virology*. doi:10.1128/jvi.01200-09.

Hurt, A. C., and I. G. Barr. 2008. "Influenza Viruses with Reduced Sensitivity to the Neuraminidase Inhibitor Drugs in Untreated Young Children." *Commun Dis Intell*.

Isai, Alina, Julie Durand, Steven Le Meur, Ana Hidalgo-Simon, and Xavier Kurz. 2012. "Autoimmune Disorders after Immunisation with Influenza A/H1N1 Vaccines with and without Adjuvant: EudraVigilance Data and Literature Review." *Vaccine*. doi:10.1016/j.vaccine.2012.09.032.

Jain, A. 2017. "Computer Aided Drug Design." In *Journal of Physics: Conference Series*. doi:10.1088/1742-6596/884/1/012072.

Jalily, P. H., J. Eldstrom, S. C. Miller, D. C. Kwan, S. S. - H. Tai, D. Chou, M. Niikura, I. Tietjen, and D. Fedida. 2016. "Mechanisms of Action of Novel Influenza A/M2 Viroporin Inhibitors Derived from Hexamethylene Amiloride." *Molecular Pharmacology*. doi:10.1124/mol.115.102731.

Kapetanovic, I. M. 2008. "Computer-Aided Drug Discovery and Development (CADDD): In Silico-Chemico-Biological Approach." *Chemico-Biological Interactions.* doi:10.1016/j.cbi.2006.12.006.

Krumbholz, Andi, Michaela Schmidtke, Silke Bergmann, Susann Motzke, Katja Bauer, Jürgen Stech, Ralf Dürrwald, Peter Wutzler, and Roland Zell. 2009. "High Prevalence of Amantadine Resistance among Circulating European Porcine in Fluenza A Viruses." *Journal of General Virology.* doi:10.1099/vir.2008.007260-0.

Leang, Sook Kwan, Simon Kwok, Sheena G. Sullivan, Sebastian Maurer-Stroh, Anne Kelso, Ian G. Barr, and Aeron C. Hurt. 2014. "Peramivir and Laninamivir Susceptibility of Circulating Influenza A and B Viruses." *Influenza and Other Respiratory Viruses.* doi:10.1111/irv.12187.

Mckimm-Breschkin, Jennifer L. 2013. "Influenza Neuraminidase Inhibitors: Antiviral Action and Mechanisms of Resistance." *Influenza and Other Respiratory Viruses.* doi:10.1111/irv.12047.

McKimm-Breschkin, Jennifer L., and Susan Barrett. 2015. "Neuraminidase Mutations Conferring Resistance to Laninamivir Lead to Faster Drug Binding and Dissociation." *Antiviral Research.* doi:10.1016/j.antiviral.2014.12.004.

Mondal, Debasis. 2011. "Zanamivir." In *XPharm: The Comprehensive Pharmacology Reference.* doi:10.1016/B978-008055232-3.62875-2.

Moscona, Anne. 2005. "Oseltamivir Resistance — Disabling Our Influenza Defenses." *New England Journal of Medicine.* doi:10.1056/nejmp058291.

Muratore, G., L. Goracci, B. Mercorelli, A. Foeglein, P. Digard, G. Cruciani, G. Palu, and A. Loregian. 2012. "Small Molecule Inhibitors of Influenza A and B Viruses That Act by Disrupting Subunit Interactions of the Viral Polymerase." *Proceedings of the National Academy of Sciences.* doi:10.1073/pnas.1119817109.

Nieva, José Luis, Vanesa Madan, and Luis Carrasco. 2012. "Viroporins: Structure and Biological Functions." *Nature Reviews Microbiology.* doi:10.1038/nrmicro2820.

Noshi, Takeshi, Mitsutaka Kitano, Keiichi Taniguchi, Atsuko Yamamoto, Shinya Omoto, Keiko Baba, Takashi Hashimoto, et al. 2018. "In Vitro Characterization of Baloxavir Acid, a First-in-Class Cap-Dependent Endonuclease Inhibitor of the Influenza Virus Polymerase PA Subunit." *Antiviral Research.* doi:10.1016/j.antiviral.2018.10.008.

Perez-Padilla, Rogelio, Daniela de la Rosa-Zamboni, Samuel Ponce de Leon, Mauricio Hernandez, Francisco Quiñones-Falconi, Edgar Bautista, Alejandra Ramirez-Venegas, et al. 2009. "Pneumonia and Respiratory Failure from Swine-Origin Influenza A (H1N1) in Mexico." *New England Journal of Medicine.* doi:10.1056/NEJMoa0904252.

Pielak, Rafal M., and James J. Chou. 2011. "Influenza M2 Proton Channels." *Biochimica et Biophysica Acta - Biomembranes.* doi:10.1016/j.bbamem.2010.04.015.

Pinto, Lawrence H., and Robert A. Lamb. 2006. "Influenza Virus Proton Channels." *Photochemical and Photobiological Sciences.* doi:10.1039/b517734k.

Pizzorno, Andrés, Yacine Abed, Chantal Rhéaume, and Guy Boivin. 2014. "Oseltamivir-Zanamivir Combination Therapy Is Not Superior to Zanamivir Monotherapy in Mice Infected with Influenza A(H3N2) and A(H1N1)Pdm09 Viruses." *Antiviral Research.* doi:10.1016/j.antiviral.2014.02.017.

Poland, Gregory A., Robert M. Jacobson, and Inna G. Ovsyannikova. 2009. "Influenza Virus Resistance to Antiviral Agents: A Plea for Rational Use." *Clinical Infectious Diseases.* doi:10.1086/598989.

Premkumar, A., L. Wilson, G. D. Ewart, and P. W. Gage. 2004. "Cation-Selective Ion Channels Formed by P7 of Hepatitis C Virus Are Blocked by Hexamethylene Amiloride." *FEBS Letters.* doi:10.1016/S0014-5793(03)01453-4.

Pushpakom, Sudeep, Francesco Iorio, Patrick A. Eyers, K. Jane Escott, Shirley Hopper, Andrew Wells, Andrew Doig, et al. 2018. "Drug Repurposing: Progress, Challenges and Recommendations." *Nature Reviews Drug Discovery.* doi:10.1038/nrd.2018.168.

Radosevic, Draginja, Milan Sencanski, Vladimir Perovic, Nevena Veljkovic, Jelena Prljic, Veljko Veljkovic, Emily Mantlo, Natalya

Bukreyeva, Slobodan Paessler, and Sanja Glisic. 2019. "Virtual Screen for Repurposing of Drugs for Candidate Influenza a M2 Ion-Channel Inhibitors." *Frontiers in Cellular and Infection Microbiology* 9 (March): 1–9. doi:10.3389/fcimb.2019.00067.

Rowe, Thomas, David Banner, Amber Farooqui, Derek C.K. Ng, Alyson A. Kelvin, Salvatore Rubino, Stephen Shih Hsien Huang, Yuan Fang, and David J. Kelvin. 2010. "In Vivo Ribavirin Activity against Severe Pandemic H1N1 Influenza A/Mexico/4108/2009." *Journal of General Virology*. doi:10.1099/vir.0.024323-0.

Saito, T., G. Taylor, and R. G. Webster. 1995. "Steps in Maturation of Influenza A Virus Neuraminidase." *Journal of Virology*.

Sakaguchi, T., Q. Tu, L. H. Pinto, and R. A. Lamb. 2002. "The Active Oligomeric State of the Minimalistic Influenza Virus M2 Ion Channel Is a Tetramer." *Proceedings of the National Academy of Sciences*. doi:10.1073/pnas.94.10.5000.

Salmaso, Veronica, and Stefano Moro. 2018. "Bridging Molecular Docking to Molecular Dynamics in Exploring Ligand-Protein Recognition Process: An Overview." *Frontiers in Pharmacology*. doi:10.3389/fphar.2018.00923.

Schnell, Jason R., and James J. Chou. 2008. "Structure and Mechanism of the M2 Proton Channel of Influenza A Virus." *Nature*. doi:10.1038/nature06531.

Shaw, Megan L;, and Peter Palese. 2013. "Orthomyxoviridae: The Viruses and Their Replication." In *Fields Virology*.

Shen, Zuyuan, Kaiyan Lou, and Wei Wang. 2015. "New Small-Molecule Drug Design Strategies for Fighting Resistant Influenza A." *Acta Pharmaceutica Sinica B*. doi:10.1016/j.apsb.2015.07.006.

Shetty, Avinash K., and Leigh A. Peek. 2012. "Peramivir for the Treatment of Influenza." *Expert Review of Anti-Infective Therapy*. doi:10.1586/eri.11.174.

Shobugawa, Yugo, Reiko Saito, Isamu Sato, Takashi Kawashima, Clyde Dapat, Isolde Caperig Dapat, Hiroki Kondo, Yasushi Suzuki, Kousuke Saito, and Hiroshi Suzuki. 2012. "Clinical Effectiveness of Neuraminidase Inhibitors - Oseltamivir, Zanamivir, Laninamivir, and

Peramivir - For Treatment of Influenza A(H3N2) and A(H1N1)Pdm09 Infection: An Observational Study in the 2010-2011 Influenza Season in Japan." *Journal of Infection and Chemotherapy.* doi:10.1007/s10156-012-0428-1.

Shtyrya, Y. A., L. V. Mochalova, and N. V. Bovin. 2009. "Influenza Virus Neuraminidase: Structure and Function." *Acta Naturae.*

Smith, Thomas F., E. Omer Burgert, Walter R. Dowdle, Gary R. Noble, R. Jean Campbell, and Robert E. Van Scoy. 2010. "Isolation of Swine Influenza Virus from Autopsy Lung Tissue of Man." *New England Journal of Medicine.* doi:10.1056/nejm197603252941308.

Tambunan. 2013. "In Silico Design of the M2 Proton Channel Inhibitors of H1N1 Virus." *OnLine Journal of Biological Sciences* 13 (1): 1–12. doi:10.3844/ojbsci.2013.1.12.

Tambunan, U. S.F., R. Harganingtyas, and A. A. Parikesit. 2012. "In Silico Modification of (1R, 2R, 3R, 5S)-(-)- Isopinocampheylamine as Inhibitors of M2 Proton Channel in Influenza a Virus Subtype H1N1, Using the Molecular Docking Approach." *Trends in Bioinformatics* 5 (2): 25–46. doi:10.3923/tb.2012.25.46.

Tambunan, Usman Sumo Friend, Arli Aditya Parikesit, Yonaniko Dephinto, and Feimmy Ruth Pratiwi Sipahutar. 2014. "Computational Design of Drug Candidates for Influenza A Virus Subtype H1N1 by Inhibiting the Viral Neuraminidase-1 Enzyme." *Acta Pharmaceutica* 64 (2): 157–72. doi:10.2478/acph-2014-0015.

Tambunan, Usman Sumo Friend, rizky archintya Rachmania, and Arli Aditya Parikesit. 2015. "In Silico Modification of Oseltamivir as Neuraminidase Inhibitor of Influenza A Virus Subtype H1N1." *Journal of Biomedical Research* 29: 150–59. doi:10.7555/JBR.29.20130024.

Tintori, Cristina, Ilaria Laurenzana, Anna Lucia Fallacara, Ulrich Kessler, Beatrice Pilger, Lilli Stergiou, and Maurizio Botta. 2014. "High-Throughput Docking for the Identification of New Influenza A Virus Polymerase Inhibitors Targeting the PA-PB1 Protein-Protein Interaction." *Bioorganic and Medicinal Chemistry Letters.* doi:10.1016/j.bmcl.2013.11.019.

Tomozawa, Takanori, Kazuki Hoshino, Makoto Yamashita, and Shuku Kubo. 2019. "Efficacy of Laninamivir Octanoate in Mice with Advanced Inflammation Stage Caused by Infection of Highly Lethal Influenza Virus." *Journal of Infection and Chemotherapy*, no. xxxx: 7–11. doi:10.1016/j.jiac.2019.02.023.

Usman Sumo Friend Tambunan. 2012. "In Silico Design of Cyclic Peptides as Influenza Virus, a Subtype H1N1 Neuraminidase Inhibitor." *African Journal of Biotechnology*. doi:10.5897/AJB11.4094.

Varghese, J. N., and P. M. Colman. 1991. "Three-Dimensional Structure of the Neuraminidase of Influenza Virus A/Tokyo/3/67 at 2·2 Å Resolution." *Journal of Molecular Biology*. doi:10.1016/0022-2836(91) 80068-6.

Vasin, A. V., O. A. Temkina, V. V. Egorov, S. A. Klotchenko, M. A. Plotnikova, and O. I. Kiselev. 2014. "Molecular Mechanisms Enhancing the Proteome of Influenza A Viruses: An Overview of Recently Discovered Proteins." *Virus Research*. doi:10.1016/j.virusres. 2014.03.015.

Veljkovic, Nevena, Sanja Glisic, Jelena Prljic, Vladimir Perovic, and Veljko Veljkovic. 2013. "Simple and General Criterion for 'in Silico' Screening of Candidate HIV Drugs." *Current Pharmaceutical Biotechnology*. doi:10.2174/13892010140513111105301.

Vellozzi, Claudia, Shahed Iqbal, and Karen Broder. 2014. "Guillain-Barré Syndrome, Influenza, and Influenza Vaccination: The Epidemiologic Evidence." *Clinical Infectious Diseases*. doi:10.1093/cid/ciu005.

Venkataraman, Padmavati, Robert A. Lamb, and Lawrence H. Pinto. 2005. "Chemical Rescue of Histidine Selectivity Filter Mutants of the M2 Ion Channel of Influenza A Virus." *Journal of Biological Chemistry*. doi:10.1074/jbc.M412406200.

Vlachakis, Dimitrios, Elena Bencurova, Nikitas Papangelopoulos, and Sophia Kossida. 2014. "Current State-of-the-Art Molecular Dynamics Methods and Applications." In *Advances in Protein Chemistry and Structural Biology*. doi:10.1016/B978-0-12-800168-4.00007-X.

Wang, Jun, Yibing Wu, Chunlong Ma, Giacomo Fiorin, Jizhou Wang, Lawrence H. Pinto, Robert A. Lamb, Michael L. Klein, and William F.

DeGrado. 2013. "Structure and Inhibition of the Drug-Resistant S31N Mutant of the M2 Ion Channel of Influenza A Virus." *Proceedings of the National Academy of Sciences.* doi:10.1073/pnas.1216526110.

Wang, Kuanglei, Fei Yang, Lihui Wang, Kemin Liu, Lu Sun, Bin Lin, Yaping Hu, Boyu Wang, Maosheng Cheng, and Yongshou Tian. 2017. "Synthesis and Biological Evaluation of NH2-Acyl Oseltamivir Analogues as Potent Neuraminidase Inhibitors." *European Journal of Medicinal Chemistry.* doi:10.1016/j.ejmech.2017.10.004.

Wang, Zhen, Li Ping Cheng, Xing Hua Zhang, Wan Pang, Liang Li, and jin long Zhao. 2017. "Design, Synthesis and Biological Evaluation of Novel Oseltamivir Derivatives as Potent Neuraminidase Inhibitors." *Bioorganic and Medicinal Chemistry Letters* 27 (23–24). Elsevier Ltd: 5429–35. doi:10.1016/j.bmcl.2018.10.040.

Xue, Hanqing, Jie Li, Haozhe Xie, and Yadong Wang. 2018. "Review of Drug Repositioning Approaches and Resources." *International Journal of Biological Sciences.* doi:10.7150/ijbs.24612.

Yamashita, Makoto. 2010. "Laninamivir and Its Prodrug, CS-8958: Long-Acting Neuraminidase Inhibitors for the Treatment of Influenza." *Antiviral Chemistry and Chemotherapy.* doi:10.3851/IMP1688.

Yang, Tianrui. 2019. "Baloxavir Marboxil: The First Cap-Dependent Endonuclease Inhibitor for the Treatment of Influenza." *Annals of Pharmacotherapy.* doi:10.1177/1060028019826565.

Yu, Wenbo, and Alexander D. Mackerell. 2017. "Computer-Aided Drug Design Methods." In *Methods in Molecular Biology.* doi:10.1007/978-1-4939-6634-9_5.

Zebedee, S L, and R A Lamb. 1988. "Influenza A Virus M2 Protein: Monoclonal Antibody Restriction of Virus Growth and Detection of M2 in Virions." *Journal of Virology.*

Zhao, Xin, Chufang Li, Shaogao Zeng, and Wenhui Hu. 2011. "Discovery of Highly Potent Agents against Influenza A Virus." *European Journal of Medicinal Chemistry.* doi:10.1016/j.ejmech.2010.10.010.

Zhao, Zhi Xiang, Li Ping Cheng, Meng Li, Wan Pang, and Fan Hong Wu. 2019. "Discovery of Novel Acylhydrazone Neuraminidase Inhibitors." *European Journal of Medicinal Chemistry*. Elsevier Masson SAS. doi:10.1016/j.ejmech.2019.04.006.

In: Advances in Medicine and Biology   ISBN: 978-1-53616-181-6
Editor: Leon V. Berhardt   © 2019 Nova Science Publishers, Inc.

Chapter 5

HISTONE ACETYLATION IN ACUTE KIDNEY INJURY (AKI)

Saeed Azimi, Ehsan Dordizadeh Basirabad and Hadi Esmaeeli

Faculty of Pharmacy, Mazandaran University of Medical Sciences,
Sari, Iran
Faculty of Medical Sciences, Mazandaran University
of Medical Sciences, Sari, Iran
R&D Department, NIAK Pharmaceuticals Co., Gorgan, Iran

ABSTRACT

Acute kidney injury (AKI) is a life-threatening situation which has a mortality rate of up to 50 percent. DNA damage, oxidative stress, apoptosis of tubular epithelial cells, and some other cell-mediated processes are the most common mechanisms involved in AKI. Histone deacetylase (HDAC) inhibitors are a new group of medications which were first used for treating different types of cancers all around the world. However, recent in vitro studies of HDAC inhibitors demonstrated significant immunomodulatory and anti-apoptotic effects of this group in renal cells which could be beneficial in the management of AKI. Since preventing such a dangerous situation seems vital, this article reviews the effects of HDAC inhibitors in preventing AKI.

The studies on the effect of inhibition of HDAC1 gene expression in renal interstitial fibroblasts and tubular epithelial cells indicate that HDAC1 enzyme has a huge role in the activation of myofibroblasts and epithelial cells. Moreover, HDAC inhibitors can prevent TGFβ-1 induced fibrogenesis and can exert anti-fibrotic effects in the kidney. Down-regulation of collagen gene expression by acetylation of histones causes an increase in expression of inhibitors of DNA binding/differentiation 2 (Id2), Bone morphologic protein 7 (BMP-7), and E-cadherin; and this process causes a significant improvement in the kidney, bone, and brown adipose tissues. Also, HDAC2 gene suppression decreases the expression of fibronectin and α-SMA. In general, HDAC inhibitors not only can suppress pathologic genes but also have various positive effects on damaged tissues in comparison to healthy tissues.

There is various evidence showing that HDAC inhibitors can have preventive and even therapeutic effects in AKI patients. Since preventing AKI seems vital, this review article focuses on the effects of HDAC inhibitors in preventing AKI.

Keywords: histone, acute kidney injury, histone deacetylase, anti-fibrotic effects

INTRODUCTION

Acute kidney injury, formerly known as acute renal failure, is a condition characterized by sudden impairment of kidney function resulting in a rapid decrease of glomerular filtration rate and elevation of serum creatinine and other nitrogenous products (Levey and James 2017). The condition frequently occurs among hospitalized patients, especially in critically conditioned cases. It is still considered one of the most challenging conditions and is associated with poor prognosis (Bellomo, Kellum, and Ronco 2012, Koza 2016). The incidence of AKI has increased over the past few years. This elevation in incidence may be due to an aging population, increased prevalence of diseases that might lead to AKI, improvements in diagnostic methods, and utilizing more sensitive diagnostic criteria (Wang et al. 2016). The incidence of AKI in the hospitalized population studied according to KDIGO-equivalent criteria is 19.4% in Eastern Asia, 7.5% in Southern Asia, 31.0% in Southeastern Asia, 9.0% in Central Asia, and 16.7% in Western Asia (Yang 2016). Moreover, Recent studies in the United States and Spain have shown that the incidence of AKI on average is 23.8 cases per 1000 discharges. AKI patients in South and Southeastern Asia are younger than in East Asia and Western countries and are associated with lesser comorbid conditions (Bouchard and Mehta 2016). Recent studies in the United States and Spain have shown incidences varying between an average of 23.8 cases per 1000 discharges. It is recognized that the epidemiology of AKI in developing countries differs from that of the developed world in several ways. For example, in developed regions, elderly patients predominate, whereas, in developing countries, AKI is a disease of the young and children (Cerdá et al. 2008). AKI has been growing at a rate of 14% per year since 2001. However, the in-hospital mortality associated with AKI has been on the decline starting with 21.9% in 2001 to 9.1% in 2011, although the number of AKI-related in-hospital deaths increased almost twofold from 147,943 to 285,768 deaths (Pavkov, Harding, and Burrows 2018). The disease burden of AKI results in an estimated $10 billion in additional costs to the health care system in the United States (Basile, Anderson, and Sutton 2012).

The diagnostic criteria for AKI are based on serum creatinine level and urine output. Standardized criteria, such as KDIGO (Kidney Disease Improving Global Outcomes) criteria, allow for uniform application of guidelines and consistent estimates of incidence and results (Table 2) (Kellum 2015).

AKI generally occurs due to three major reasons: 1) Diseases that cause renal hypoperfusion, resulting in decreased function without frank parenchymal damage- prerenal AKI, 2) Diseases that directly involve the renal parenchyma- intrinsic AKI, 3) Conditions related with obstruction in structures after the kidney such as urethra, urinary bladder or ureter-postrenal AKI which is generally a less common cause of AKI (Nagamani et al. 2015).

Table 1. The diagnostic criteria for AKI

Criteria and staging for acute kidney injury		
Stage	Serum Creatinine	Urine Output
1	1.5–1.9 times the baseline OR ≥0.3 mg/dL (>26.5mmol/L) increase	<0.5 mL/kg/h for 6–12 h
2	2.0–2.9 times baseline	<0.5 mL/kg/h for ≥12 h
3	3.0 times the baseline OR Increase in serum creatinine to ≥ 4.0 mg/dL (353.6mmol/L) OR Initiation of renal replacement therapy OR In patients <18 y, decrease in eGFR to <35 mL/min per 1.73 m	<0.3 mL/kg/h for ≥24 h OR Anuria for ≥12 h

Table 2. Contrast agents in renal cells damages

Cellular effects	Hypoxia and vasoconstriction	Tubular effects
- Direct cell membrane damage - Perturbation of mitochondrial function - Generation of reactive oxygen species - Apoptosis	- Constriction of afferent arterioles and/or vasa recta - Enhanced renal vascular responsiveness to angiotensin II and endothelin-1 - Endothelial damage with subsequent vasoconstriction - Increased vascular resistance by congestion - Acute hypotension (anaphylaxis)	- Perturbed tubuloglomerular feedback - Cytotoxic effects - Tubulovascular crosstalk with subsequent vasoconstriction - Tubular obstruction by increased fluid viscosity

The kidneys receive approximately one-fourth of the blood pumped by the heart at every beat, so any disturbance in general circulation or isolated disorders of the kidneys' internal circulation can have a profound effect on renal perfusion and as a result on renal function. Post-renal AKI which generally is a less common cause of AKI, obstruction of the urethra, urinary bladder or ureter increases intratubular pressure, resulting in a decreased GFR. Intrinsic AKI depending on the injured structure within the kidney, can be divided to 4 groups: 1) Tubular damage, 2) Glomerular damage, 3)

Interstitial damage, and 4) Vascular damage (Basile, Anderson, and Sutton 2012).

Intrinsic AKI is the most common cause of AKI and it may happen because of sepsis, ischemia and nephrotoxic exposure. Studies on sepsis as one of the most common etiologies of acute kidney injury (Majumdar 2010) have shown that the mechanism of sepsis-induced AKI is different from non-septic forms (Wan et al. 2008). Its pathophysiology remains relatively uncertain, but the absence of histologic changes in animal models suggest that septic AKI may be a functional phenomenon with combined microvascular shunting and tubular cell stress (Bellomo et al. 2017).

Ischemia is another common etiology of AKI (Xu et al. 2016) and the main mechanisms involved in this process are hemodynamic changes, inflammation, and endothelial and epithelial cell injury. These events are followed by repair that can be either adaptive repair and lead to the regaining of the epithelial cell integrity, or maladaptive repair and cause chronic kidney disease (CKD) (Bonventre and Yang 2011). Inflammation appears to be the mutual factor that connects the various cell types involved in this procedure (Sharfuddin and Molitoris 2011). An extreme and continuous renal vasoconstriction that reduces overall renal blood flow to approximately half of the normal amount has been considered a hallmark of intrinsic AKI. Low level of oxygen results in rapid degradation of ATP to ADP and AMP. With prolonged ischemia, AMP is metabolized further to hypoxanthine and adenine nucleotide. Hypoxanthine accumulation contributes to the production of reactive oxygen molecule while Adenine nucleotides freely diffuse out of cells and their depletion prohibits re-synthesis of intracellular ATP during reperfusion. Ischemia also leads to early disturbance of at least two basolaterally polarized proteins, namely Na/K-ATPase and integrins.

Increasing evidence now points out that apoptosis is the main mechanism of early tubule cell death in current clinical AKI (Devarajan 2006).

Epidemiologic statistics show that a significant part of AKI cases is at least partially connected to nephrotoxin exposure (Ferguson, Vaidya, and Bonventre 2008). Acute tubular necrosis is the most frequent form of renal injury caused by nephrotoxin exposure, although other forms of renal failure may be seen. (Pannu and Nadim 2008).

One of the most common nephrotoxins is iodinated contrast agents. Contrast-induced acute kidney injury (CIAKI) occurs in up to 30% of patients who receive iodinated contrast media and is generally considered to be the third most common cause of hospital-acquired AKI (Fahling et al. 2017). Contrast agents can damage renal cells through three major mechanisms which are summarized in Table 1 (Fahling et al. 2017).

The assessment and initial management of patients with acute kidney injury (AKI) should include: (1) an evaluation of the contributing factors of the injury, (2) an evaluation of the clinical course including comorbidities, (3) a careful evaluation of volume status, and (4) starting appropriate therapeutic actions designed to reverse or stop worsening of situation (Himmelfarb et al. 2008).

Epigenetic is the science of reversible heritable changes in the function and expression of genes without altering their nucleotide sequences. In other words, changes occur in the phenotype of the cells independent of DNA sequences (Nicoglou and Merlin 2017). In 1939, the term "epigenetics" was used for the first time to explain the formation of different phenotypes due to the interactions between respective genes and their products (Javaid and Choi 2017, Waddington 1939). In epigenetics, events around chromatin and DNA are described. Various modifications occur in the form of epigenetic processes within different parts of the DNA which can change DNA structure (Dawson and Kouzarides 2012). In the eukaryotic cells, DNA should be tightly packed in the core of the cell. This is accomplished through interaction between the histone protein complexes and the formation of a structure called chromatin. The nucleosome is the fundamental repeating unit of chromatin and is formed of octamers containing four nuclear histone proteins (H2A, H2B, H3, and H4) and two rounds of DNA twisted around the outer surface of the octamer. The degree of this folding directly affects some of the most important cellular tasks such as replication, transcription, and recombination (Luger et al. 1997).

Modifications and epigenetic processes within the cell can include DNA methylation, histone modifications, and etc. (Hartman and Czyz 2015).

DNA methylation is seen in various physiological processes of the cell, such as chromosome deactivation, gene expression, aging, and cancer. It can also play a role in the structure of chromatin, its stability, and compatibility. In addition, it can participate in the process of the tumor cell cycle, gene transcription, DNA damage repair, cell differentiation and the metabolism of anticancer drugs (Song et al. 2015). In mammals, three known proteins are characterized to regulate DNA methylation (DNMT1, DNMT3a, and DNMT3b) which are expressed in human tumor cells (Sang and Deng 2019). Methylation is done in several ways including mono, di, and trimethylation of lysine and mono and dimethylation of arginine residues (Felsenfeld and Groudine 2003). The state of methylation of the genome is mainly preserved by DNMT1 and it plays an important role in inhibiting the expression of tumor suppressor genes in human cancer cells. Moreover, reduction in the expression of DNMT3a can delay the metastasis and growth of tumor cells in some cancers like melanoma (Sang and Deng 2019).

Histone modifications are one of the remarkable epigenetic ways in the cells with certain physiological and biochemical properties (Day 2014). The histone tail is linked to DNA by H1 histone and most post-translational changes of histones occur in 15 to 38 amino acid residues in this domain (Furumatsu and Asahara 2010). In addition, the histone tail plays a key role in the process of chromatin accumulation. The degree of chromatin density directly affects DNA replication, recombination, and transcription. Histone changes can alter the structure and function of chromatin, which can affect the transcription of many key genes, and plays an important role in the development of tumors (Khan, Reddy, and Gupta 2015).

Histones can be modified in many ways. There are more than 60 histone tails that can be changed. N-terminal amino acid residues of histones are post-transplantation sites. Histones modifications are carried out for various purposes, such as DNA repair, regulating transcription processes, and cell replication by means of several enzymes (Bannister and Kouzarides 2011). These modifications include acetylation, methylation, phosphorylation, ubiquitylation, sumolytion, ADP ribosylation, deminition, and proline isomerization.

Currently, the most studied modifications are methylation and acetylation. Histone methyltransferases regulate the cell cycle. Abnormal gene expression and excessive expression of histone methyltransferases exist in many tumors, while histone acetylation plays an important role in many of the cells such as nucleosomes assembly (Kouzarides 2007).

Histone acetylases are another group of enzymes that cause chromatin modification and affect gene expression by acetylation. These enzymes transfer an acetylchoenzyme A group to the lysine amino acid in the N-terminal domain of histone. Lysine acetylation leads to neutralizing the positive charge of DNA and weakening the interactions between nucleosomes and histones. This process can make changes in the configuration and structure of DNA and give the opportunity to other enzymatic complexes to have more access to a particular part of the genome.

The acetylation can be reversed by histone deacetylases (Sterner and Berger 2000). In human, 18 histone deacetylase enzymes have been discovered which use either NAD^+ or zinc to deacetylase lysine amino acid in histone or other proteins. Deacetylation of non-histone proteins affects various intracellular processes, while histone deacetylation has a key role in DNA-related activities. Therefore, histone deacetylase inhibitors are now used for anti-cancer purposes (Seto and Yoshida 2014).

The histone deacetylase (HDAC) family consists of three classes of proteins: class I, II, and IV. Proteins in each class have an almost similar three-dimensional structure, function, and sequence. Class I includes HDAC1, HDAC2, HDAC3, and HDAC8 which are similar to HAD1 yeast protein. Class II enzymes are divided into two subclasses: HDAC4, HDAC5, HDAC, HDAC9 in subclass IIa and HDCA6 and HDCA10 in subclass IIb. HDAC11 has similarities to both IIa and IIb subclasses. The class IV is different in some features with the other two classes. These enzymes (also called SIRT instead of HDAC) are SIRT2, SIRT3, SIRT4, SIRT5, SIRT6, and SIRT7. SIRT enzymes use NAD^+ as co-factor (Peng, Yuan, and Seto 2015, Habibi and Esmaeeli 2017).

METHODS

In this review article, valid scientific articles indexed in literature review electronic databases such as "Web of Science, Scopus, PubMed" were collected. The used keywords in this review are "Histone," "Acute Kidney Injury," "Histone Deacetylase" and "Anti-fibrotic effects".

DISCUSSION

Acute kidney injury (AKI) is a syndrome characterized by a sudden decrease in urinary excretion. As a result, there will be a failure in the maintenance of electrolyte, fluids, and acid-base balance. Some of the signs and symptoms of this syndrome include oliguria, the rise of creatinine and an increase in urea, phosphate, and potassium concentration. The definition of AKI provided by KDIGO (Kidney Disease Improving Global Outcomes) is as follows: Increase in serum creatinine equal to or greater than 0.3 mg/dl over 48 hours, a change in serum creatinine equal to or more than 1.5 times the baseline, assuming that it occurred about seven days ago or urine volume less than 0.5 ml/kg/h in 6 hours. This syndrome has a high prevalence and high morbidity and mortality (Rodríguez-Romo et al. 2015, Kellum and Lameire 2013).

Table 3. Current evidence for a functional role of extracellular histones in kidney

Disorder	Pathomechanism	References
Acute kidney injury	Microvascular endothelial injury and TLR2/4-mediated inflammation leading to acute tubular necrosis	(Rosin and Okusa 2012) (Allam et al. 2012)
Glomerulonephritis	NETosis leading to vascular injury in glomeruli	(Kessenbrock et al. 2009)
Circulatory death kidneys	Quantitation of extracellular histones leading to the assessment of posttransplant graft function and survival	(van Smaalen et al. 2017)

Epigenetics has been studied in various diseases and AKI is one of them (Ataee and Esmaeeli 2017). Chromatin compression, DNA methylation, acetylation, and DNA deacetylation are the most studied subjects in AKI. These modifier mechanisms increase the risk of producing pro-inflammatory and prophylactic cytokines such as complement protein 3 (C3) and transforming growth factor β (TGF-β). This process ultimately causes the conversion of the epithelial cells to the mesenchymal cells, resulting in kidney fibrosis (Rodríguez-Romo et al. 2015). Electrostatic interaction of the positive charge on histones and the negative charge on DNA are related to the condensed chromatin structure (Bomsztyk and Denisenko 2013a). Acetylation of histones removes the positive charges, thereby reducing the affinity of histones for the charged DNA, subsequently altering the condensed chromatin to a relaxed structure for activation of activators or inhibitors of gene transcription. HDAC activity can reverse this situation (Tang and Zhuang 2015).

During AKI, Tubular epithelial cells are exposed to hypoxia, which not only modifies metabolism but also affects the structure of chromatin, as well as the binding of various transcription factors (Bomsztyk and Denisenko 2013b). It has been reported that an increase in the expression of pro-inflammatory cytokines such as tumor necrosis factor alfa (TNF-α) and monocyte chemoattractant protein-1 (MCP-1) is seen after AKI (Zager and Johnson 2009).

In hypoxic conditions, cholesterol is produced in cells having a protective role in cell membrane stabilization and mitochondrial activity. It was reported in a study that the production of RNA polymerase II, HMG-CoA reductase and some proteins like sterol regulatory element binding protein-1 and 2 (SREBP-1 and SREBP-2) increases, leading to the synthesis of cholesterol. This protective process occurs after trimethylation of histone 3-lysine 4 and acetylation of histone 3-lysine 9. Inhibition of each mechanism may develop renal injury and maybe that is why in some other diseases, HDAC inhibitors (HDACi) are used to exert anti-inflammatory and anti-fibrotic effects (Kazantsev and Thompson 2008, Naito, Bomsztyk, and Zager 2009).

In some researches, the efficacy of HDAC inhibitors has been studied on AKI. Following severe unilateral renal ischemia/reperfusion (I/R), a transient hypoacetylation in Proximal tubular cells of mice was seen which was recovered after 24 hours because of the decline in HDAC5 levels. Removing HADC5 significantly increased the histone acetylation and increased BMP7, which helps to heal tubular epithelium. Using a unilateral ureteral obstruction model (UUO), the roles of HDACs were investigated again. It turns out that activated HDAC1 and HDAC2 are responsible for hypoacetylation in injured Kidney. Also, levels of HDAC7 was decreased in this condition (Marumo et al. 2008, Marumo et al. 2010). HDAC6 inhibition suppressed oxidative stress by reducing the level of malondialdehyde and increasing the expression of superoxide dismutase in AKI (Shi et al. 2017). It can be understood that some of HDACs have protective roles as it has been reported that there is a relation between high levels of SIRT1 and improvement of AKI because SIRT1 is responsible for protection in some stress-related pathways (Fan et al. 2013). But it is possible that lack of regulation of acetylation of histones and an increase in the levels of inflammation may develop AKI (Isbir et al. 2007, Perianayagam et al. 2007). SIRT3 is the physiological histone deacetylase that inhibits p300-mediated histone acetylation (Wang et al. 2012). Epigenetic modification has also been demonstrated to endotoxin hyper-responsiveness of cytokine production can be responsible in the pathogenesis of AKI (Tang and Zhuang 2015). Some studies reported that inhibiting class I HDAC activity induced histone H3 hyperacetylation, decreased cell proliferation, and reduced expression of cyclin D1 and proliferating cell nuclear antigen in cultured renal proximal tubular cells, suggesting the effect of class I HDACs in adjusting renal tubular cell proliferation (Tang and Zhuang 2015). TSA, a HDACi, stimulated AMPK and inactivated mTOR during cisplatin treatment of proximal tubule cells and kidneys in mice and may protect kidneys by activating autophagy (Liu et al. 2018).

Autophagy is a lysosomal degradation pathway that removes protein aggregates and dysfunctional organelles. In cellular stress, autophagy is activated and serves primarily as an adaptive mechanism for cell survival, whereas deregulated autophagy plays important roles in the pathogenesis of various diseases (Liu et al. 2018). Studies showed the importance of autophagy to the protective effects of HDACi in nephrotoxicity, suggesting a role of autophagy in kidney recovery from injury (Cheng et al. 2015).

CONCLUSION

Histone acetylation plays an important role in the pathogenesis of AKI. More studies are needed to determine the specific HDAC isoforms and acetylated histones involved in AKI. Protective effects of HDACi is not dispensable in kidney recovery from injury.

REFERENCES

Allam, Ramanjaneyulu, Christina Rebecca Scherbaum, Murthy Narayana Darisipudi, Shrikant R Mulay, Holger Hägele, Julia Lichtnekert, Jan Henrik Hagemann, Khader Valli Rupanagudi, Mi Ryu, and Claudia Schwarzenberger. 2012. "Histones from dying renal cells aggravate kidney injury via TLR2 and TLR4." *Journal of the American Society of Nephrology* 23 (8):1375-1388.

Ataee, R, and H Esmaeeli. 2017. "The Renoprotective Effects of Sodium Valproate as a Histone Deacetylase Inhibitor on Diabetic Nephropathy." *Journal of Babol University of Medical Sciences* 19 (9):45-53. doi: 10.22088/jbums.19.9.45.

Bannister, Andrew J., and Tony Kouzarides. 2011. "Regulation of chromatin by histone modifications." *Cell research* 21 (3):381-395. doi: 10.1038/cr.2011.22.

Basile, D. P., M. D. Anderson, and T. A. Sutton. 2012. "Pathophysiology of acute kidney injury." *Compr Physiol* 2 (2):1303-53.

Bellomo, R., J. A. Kellum, and C. Ronco. 2012. "Acute kidney injury." *Lancet* 380 (9843):756-66.

Bellomo, Rinaldo, John A. Kellum, Claudio Ronco, Ron Wald, Johan Martensson, Matthew Maiden, Sean M. Bagshaw, Neil J. Glassford, Yugeesh Lankadeva, Suvi T. Vaara, and Antoine Schneider. 2017. "Acute kidney injury in sepsis." *Intensive Care Medicine* 43 (6):816-828. doi: 10.1007/s00134-017-4755-7.

Bomsztyk, K., and O. Denisenko. 2013a. "Epigenetic alterations in acute kidney injury." *Semin Nephrol* 33 (4):327-40. doi: 10.1016/j.semnephrol.2013.05.005.

Bomsztyk, Karol, and Oleg Denisenko. 2013b. "Epigenetic alterations in acute kidney injury." *Seminars in Nephrology* 33(4):327-340.

Bonventre, J. V., and L. Yang. 2011. "Cellular pathophysiology of ischemic acute kidney injury." *J Clin Invest* 121 (11):4210-21.

Bouchard, J., and R. L. Mehta. 2016. "Acute Kidney Injury in Western Countries." *Kidney Dis* 2 (3):103-110.

Cerdá, Jorge, Norbert Lameire, Paul Eggers, Neesh Pannu, Sigehiko Uchino, Haiyan Wang, Arvind Bagga, and Adeera Levin. 2008. "Epidemiology of Acute Kidney Injury." *Clinical Journal of the American Society of Nephrology* 3 (3):881-886. doi: 10.2215/cjn.04961107.

Cheng, Huifang, Xiaofeng Fan, William E. Lawson, Paisit Paueksakon, and Raymond C. Harris. 2015. "Telomerase deficiency delays renal recovery in mice after ischemia-reperfusion injury by impairing autophagy." *Kidney International* 88 (1):85-94. doi: 10.1038/ki.2015.69.

Dawson, Mark A, and Tony Kouzarides. 2012. "Cancer epigenetics: from mechanism to therapy." *Cell* 150 (1):12-27.

Day, Jeremy J. 2014. "New approaches to manipulating the epigenome." *Dialogues in Clinical Neuroscience* 16 (3):345-357.

Devarajan, Prasad. 2006. "Update on Mechanisms of Ischemic Acute Kidney Injury." *Journal of the American Society of Nephrology* 17 (6):1503-1520. doi: 10.1681/asn.2006010017.

Fahling, M., E. Seeliger, A. Patzak, and P. B. Persson. 2017. "Understanding and preventing contrast-induced acute kidney injury." *Nat Rev Nephrol* 13 (3):169-180.

Fan, H., H. C. Yang, L. You, Y. Y. Wang, W. J. He, and C. M. Hao. 2013. "The histone deacetylase, SIRT1, contributes to the resistance of young mice to ischemia/reperfusion-induced acute kidney injury." *Kidney Int* 83 (3):404-13. doi: 10.1038/ki.2012.394.

Felsenfeld, Gary, and Mark Groudine. 2003. "Controlling the double helix." *Nature* 421 (6921):448.

Ferguson, Michael A, Vishal S Vaidya, and Joseph V Bonventre. 2008. "Biomarkers of nephrotoxic acute kidney injury." *Toxicology* 245 (3):182-193.

Furumatsu, Takayuki, and Hiroshi Asahara. 2010. "Histone acetylation influences the activity of Sox9-related transcriptional complex." *Acta Med Okayama* 64 (6):351-357.

Habibi, E, and H Esmaeeli. 2017. "A Review of the Effects of Curcumin on Histone Acetyltransferase Activity in the Prevention of Cardiac Hypertrophy." *Journal of Babol University of Medical Sciences* 19 (1):27-35. doi: 10.22088/jbums.19.1.27.

Hartman, Mariusz L, and Malgorzata Czyz. 2015. "MITF in melanoma: mechanisms behind its expression and activity." *Cellular and Molecular Life Sciences* 72 (7):1249-1260.

Himmelfarb, Jonathan, Michael Joannidis, Bruce Molitoris, Miet Schietz, Mark D. Okusa, David Warnock, Franco Laghi, Stuart L. Goldstein, Richard Prielipp, Chirag R. Parikh, Neesh Pannu, Suzana M. Lobo, Sudhir Shah, Vincent D'Intini, and John A. Kellum. 2008. "Evaluation and Initial Management of Acute Kidney Injury." *Clinical Journal of the American Society of Nephrology* 3 (4):962-967. doi: 10.2215/cjn.04971107.

Isbir, S. C., A. Tekeli, A. Ergen, H. Yilmaz, K. Ak, A. Civelek, U. Zeybek, and S. Arsan. 2007. "Genetic polymorphisms contribute to acute kidney injury after coronary artery bypass grafting." *Heart Surg Forum* 10 (6):E439-44. doi: 10.1532/hsf98.20071117.

Javaid, Nasir, and Sangdun Choi. 2017. "Acetylation-and methylation-related epigenetic proteins in the context of their targets." *Genes* 8 (8):196.

Kazantsev, A. G., and L. M. Thompson. 2008. "Therapeutic application of histone deacetylase inhibitors for central nervous system disorders." *Nat Rev Drug Discov* 7 (10):854-68. doi: 10.1038/nrd2681.

Kellum, J. A. 2015. "Diagnostic Criteria for Acute Kidney Injury: Present and Future." *Crit Care Clin* 31 (4):621-32.

Kellum, John A, and Norbert Lameire. 2013. "Diagnosis, evaluation, and management of acute kidney injury: a KDIGO summary (Part 1)." *Critical care* 17 (1):204.

Kessenbrock, Kai, Markus Krumbholz, Ulf Schönermarck, Walter Back, Wolfgang L Gross, Zena Werb, Hermann-Josef Gröne, Volker Brinkmann, and Dieter E Jenne. 2009. "Netting neutrophils in autoimmune small-vessel vasculitis." *Nature Medicine* 15 (6):623.

Khan, Shafqat Ali, Divya Reddy, and Sanjay Gupta. 2015. "Global histone post-translational modifications and cancer: Biomarkers for diagnosis, prognosis and treatment?" *World Journal of Biological Chemistry* 6 (4):333.

Kouzarides, Tony. 2007. "Chromatin modifications and their function." *Cell* 128 (4):693-705.

Koza, Y. 2016. "Acute kidney injury: current concepts and new insights." *J Inj Violence Res* 8 (1):58-62.

Levey, A. S., and M. T. James. 2017. "Acute Kidney Injury." *Ann Intern Med* 167 (9).

Liu, Jing, Man J. Livingston, Guie Dong, Chengyuan Tang, Yunchao Su, Guangyu Wu, Xiao-Ming Yin, and Zheng Dong. 2018. "Histone deacetylase inhibitors protect against cisplatin-induced acute kidney injury by activating autophagy in proximal tubular cells." *Cell Death & Disease* 9 (3):322. doi: 10.1038/s41419-018-0374-7.

Luger, K., A. W. Mader, R. K. Richmond, D. F. Sargent, and T. J. Richmond. 1997. "Crystal structure of the nucleosome core particle at 2.8 A resolution." *Nature* 389 (6648):251-60. doi: 10.1038/38444.

Majumdar, A. 2010. "Sepsis-induced acute kidney injury." *Indian J Crit Care Med* 14 (1):14-21.

Marumo, T., K. Hishikawa, M. Yoshikawa, and T. Fujita. 2008. "Epigenetic regulation of BMP7 in the regenerative response to ischemia." *J Am Soc Nephrol* 19 (7):1311-20. doi: 10.1681/asn.2007091040.

Marumo, T., K. Hishikawa, M. Yoshikawa, J. Hirahashi, S. Kawachi, and T. Fujita. 2010. "Histone deacetylase modulates the proinflammatory and -fibrotic changes in tubulointerstitial injury." *Am J Physiol Renal Physiol* 298 (1):F133-41. doi: 10.1152/ajprenal.00400.2009.

Nagamani, R, K Sudarsi, K Amaravati, Musa Khan, P Sakuntala, and D Deepthi. 2015. "A study on clinical profile of acute kidney injury." *Ratio* 8:3.

Naito, M., K. Bomsztyk, and R. A. Zager. 2009. "Renal ischemia-induced cholesterol loading: transcription factor recruitment and chromatin remodeling along the HMG CoA reductase gene." *Am J Pathol* 174 (1):54-62. doi: 10.2353/ajpath.2009.080602.

Nicoglou, Antonine, and Francesca Merlin. 2017. "Epigenetics: A way to bridge the gap between biological fields." *Studies in History and Philosophy of Science Part C: Studies in History and Philosophy of Biological and Biomedical Sciences* 66:73-82.

Pannu, Neesh, and Mitra K. Nadim. 2008. "An overview of drug-induced acute kidney injury." *Critical Care Medicine* 36 (4):S216-S223. doi: 10.1097/CCM.0b013e318168e375.

Pavkov, Meda E, Jessica L Harding, and Nilka R Burrows. 2018. "Trends in hospitalizations for acute kidney injury—United States, 2000–2014." *Morbidity and Mortality Weekly Report* 67 (10):289.

Peng, Lirong, Zhigang Yuan, and Edward Seto. 2015. "Histone deacetylase activity assay." In *Chromatin Protocols*, 95-108. Springer.

Perianayagam, M. C., O. Liangos, A. Y. Kolyada, R. Wald, R. W. MacKinnon, L. Li, M. Rao, V. S. Balakrishnan, J. V. Bonventre, B. J. Pereira, and B. L. Jaber. 2007. "NADPH oxidase p22phox and catalase gene variants are associated with biomarkers of oxidative stress and adverse outcomes in acute renal failure." *J Am Soc Nephrol* 18 (1):255-63. doi: 10.1681/asn.2006070806.

Rodríguez-Romo, Roxana, Nathan Berman, Arturo Gómez, and Norma A Bobadilla. 2015. "Epigenetic regulation in the acute kidney injury to chronic kidney disease transition." *Nephrology* 20 (10):736-743.

Rosin, Diane L, and Mark D Okusa. 2012. "Dying cells and extracellular histones in AKI: beyond a NET effect?" J Am Soc Nephrol. 2012 Aug; 23(8):1275-127..

Sang, Yanqi, and Yu Deng. 2019. "Current insights into the epigenetic mechanisms of skin cancer." *Dermatologic Therapy* 32(4):e12964. https://doi.org/10.1111/dth.12964.

Seto, Edward, and Minoru Yoshida. 2014. "Erasers of histone acetylation: the histone deacetylase enzymes." *Cold Spring Harbor perspectives in biology* 6 (4):a018713.

Sharfuddin, A. A., and B. A. Molitoris. 2011. "Pathophysiology of ischemic acute kidney injury." *Nat Rev Nephrol* 7 (4):189-200.

Shi, Y., L. Xu, J. Tang, L. Fang, S. Ma, X. Ma, J. Nie, X. Pi, A. Qiu, S. Zhuang, and N. Liu. 2017. "Inhibition of HDAC6 protects against rhabdomyolysis-induced acute kidney injury." *Am J Physiol Renal Physiol* 312 (3):F502-f515. doi: 10.1152/ajprenal.00546.2016.

Song, Jing, Zhanwen Du, Mate Ravasz, Bohan Dong, Zhenghe Wang, and Rob M Ewing. 2015. "A protein interaction between β-catenin and dnmt1 regulates wnt signaling and dna methylation in colorectal cancer cells." *Molecular Cancer Research* 13 (6):969-981.

Sterner, David E, and Shelley L Berger. 2000. "Acetylation of histones and transcription-related factors." *Microbiol. Mol. Biol. Rev.* 64 (2):435-459.

Tang, Jinhua, and Shougang Zhuang. 2015. "Epigenetics in acute kidney injury." *Current Opinion in Nephrology and Hypertension* 24 (4):351.

van Smaalen, Tim C, Daniëlle MH Beurskens, ER Pieter Hoogland, Bjorn Winkens, Maarten HL Christiaans, Chris P Reutelingsperger, LW Ernest van Heurn, and Gerry AF Nicolaes. 2017. "Presence of cytotoxic extracellular histones in machine perfusate of donation after circulatory death kidneys." *Transplantation* 101 (4):e93-e101.

Waddington, C. H. 1939. "Preliminary Notes on the Development of the Wings in Normal and Mutant Strains of Drosophila." *Proc Natl Acad Sci U S A* 25 (7):299-307. doi: 10.1073/pnas.25.7.299.

Wan, Li, Sean M Bagshaw, Christoph Langenberg, Takao Saotome, Clive May, and Rinaldo Bellomo. 2008. "Pathophysiology of septic acute kidney injury: what do we really know?" *Critical Care Medicine* 36 (4):S198-S203.

Wang, Y., Y. Fang, J. Teng, and X. Ding. 2016. "Acute Kidney Injury Epidemiology: From Recognition to Intervention." *Contrib Nephrol* 187:1-8.

Wang, Zhiwei, Hiroyuki Inuzuka, Jiateng Zhong, Pengda Liu, Fazlul H Sarkar, Yi Sun, and Wenyi Wei. 2012. "Identification of acetylation-dependent regulatory mechanisms that govern the oncogenic functions of Skp2." *Oncotarget* 3 (11):1294.

Xu, Y., M. Guo, W. Jiang, H. Dong, Y. Han, X. F. An, and J. Zhang. 2016. "Endoplasmic reticulum stress and its effects on renal tubular cells apoptosis in ischemic acute kidney injury." *Ren Fail* 38 (5):831-7.

Yang, Li. 2016. "Acute kidney injury in Asia." *Kidney Diseases* 2 (3):95-102.

Zager, Richard A, and Ali CM Johnson. 2009. "Renal ischemia-reperfusion injury upregulates histone-modifying enzyme systems and alters histone expression at proinflammatory/profibrotic genes." *American Journal of Physiology-Renal Physiology* 296 (5):F1032-F1041.

In: Advances in Medicine and Biology   ISBN: 978-1-53616-181-6
Editor: Leon V. Berhardt   © 2019 Nova Science Publishers, Inc.

Chapter 6

MEDIAL FRACTURES OF PROXIMAL FEMUR IN YOUNG PATIENTS: A REVIEW

G. Nicolaci, A. Aprato*, E. Enrietti and A. Massè

Orthopaedics and Traumatology Department
University of Turin, Turin, Italy

ABSTRACT

Femoral neck fractures in patients under 65 years old are an uncommon injury often caused by a high-energy trauma, with possible fracture comminution and disruption of blood supply to the femoral head. Garden and Pauwels classification systems are currently used to decide the appropriate treatment; on the other hand, the literature shows controversies among surgeons' decision concerning specific treatment variables such as time to surgery, the role of capsulotomy and fixation methods. Hip arthroplasty is not necessarily the first option in young patients because it may not last as long as in low-demand patients.

The main goal in such subgroup of patients is to obtain anatomical reduction and a stable fixation to attempt femoral head preservation: it has been demonstrated that specific measurements of the femoral head posterior tilt and alignment are reliable predictors of reoperation.

* Corresponding Author's E-mail: ale_aprato@hotmail.com.

The clinical outcomes can be improved with good pre-operative planning, optimization of surgical procedures and introduction of new improved implants and techniques. Despite this, a femoral fracture in a young patient is a major adverse event: the healed femur would never be the same as the healthy one, and further surgeries might be needed over the years because of the long life-expectancy. Therefore, a fully comprehension of the anatomical and mechanical aspects of the hip appear to be essential.

This paper reviews the literature about how to obtain an optimal reduction and how to maintain it through different fixation techniques; furthermore, the functional results will be discussed in relation to the specific fixation implants, to the fracture pattern and to the patient-related factors.

Eventually, factors influencing good fracture healing and how to prevent major complications (shortening in comminuted fractures, non-union and avascular necrosis of the femoral head) will be discussed.

Keywords: hip fractures, proximal femur fracture, young patients, reduction, fixation

INTRODUCTION

The incidence of femoral neck fractures in young patients is low (3% of all the hip fractures) [1], and these are mainly associated with high-energy trauma. In the United States the frequency is higher than in Europe: more than 250000 hip fractures happen every year, comprehensive of medial and lateral femoral fractures, and 10% of them occurs in people younger than 50 years old [2].

Usually a high-energy trauma such as motor vehicle accidents and falls from height account for most femoral neck fractures in the young.

The main mechanism of injury is an axial load that acts upon an abducted and extra-rotated hip.

A minor percentage of femoral fractures in the young population is related to low energy forces that act upon comorbidities such as alcoholism and chronic disease or is caused by cyclical loading -stress as in athletes.

CLINICAL EVALUATION

Patient involved in high-energy trauma should be subjected to ATLS protocols; if they only reported a femoral neck fracture, the patients usually complain of groin and thigh pain, with the lower extremity shortened and externally rotated, and they are unable to walk. Pain is elicited with palpation, axial compression and on attempted range of motion.

Sometimes, in nondisplaced fractures derived from lower trauma, the patient might be able to weight bearing and the lower leg might lack of deformity.

All patients, especially after high energy trauma, should undergo a thorough secondary survey to evaluate for associated injuries.

RADIOGRAPHIC EVALUATION

The first radiographic evaluation should include an anteroposterior view of the pelvis and an anteroposterior and a cross-table lateral view of the involved hip; a thin slice computed tomography may help in understanding the fracture pattern and in detecting nondisplaced neck fractures [2].

CLASSIFICATION

Among the most commonly used classification system, the Garden classification is based on the degree of displacement and defined four types of fracture: type 1 is an incomplete fracture, or a complete fracture with a valgus impacted pattern; type 2 is a complete fracture without displacement of the fragments; type 3 is a complete fracture with a slight displacement of the fragments; type 4 is a complete displaced fracture [3] (Figure 1).

Pauwels classification is based on the inclination degree of the fracture line and defined three types of neck fractures: type 1 has a fracture line with an inclination under 30° from the ground plane; type 2 has an inclination degree of 30-50°; type three has a vertical fracture line, with an inclination higher than 50°.

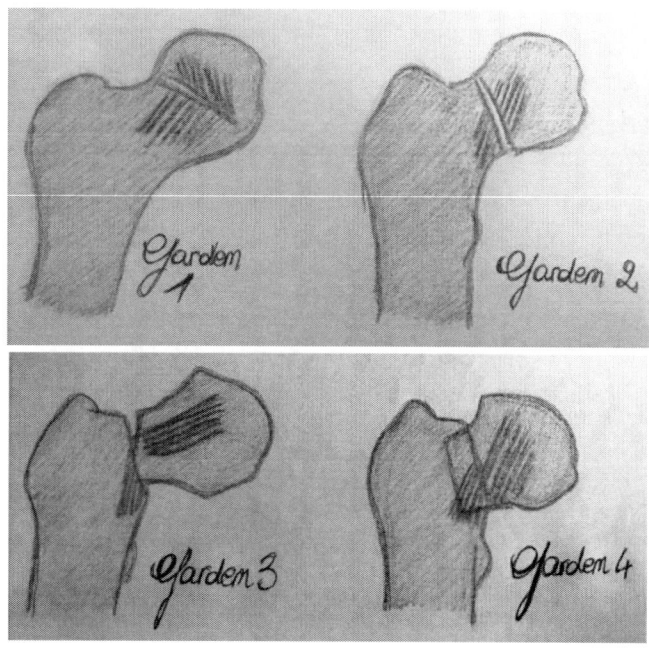

Figure 1. Garden classification.

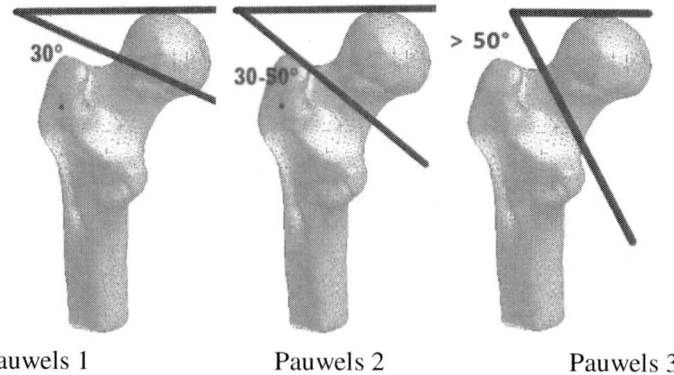

Pauwels 1   Pauwels 2   Pauwels 3

Figure 2. Pauwels classification.

Because of stronger shear forces, the increase of the inclination degree rises the risk of displacement and instability. It has been demonstrated that unstable fractures have a higher risk of avascular necrosis of the femoral head [4] (Figure 2).

Whereas Garden Classification is a powerful tool to predict the likeability of the occurrence of AVN, Pauwels Classification has not shown such reliability [5, 6].

PREOPERATIVE MANAGEMENT

Is the Time to Surgery the Keyhole?

In this group of patients the main goals to achieve are represented by preservation of the femoral head, an early mobilization and rehabilitation, a satisfactory functional result and a low complication rate.

The ideal time for reduction and fixation is not completely clarify still nowadays: the avascular necrosis of the femoral head does not rely on timing only, but it appears to be the result of complex interacting factors including the intraosseous vascularization [7].

Despite the results obtained by authors like Swiontkowski, Jain, Manninger et al. [8, 9, 10] who observed significant differences in the avascular necrosis rates within or after a specific interval time (6, 8, 12 hours after trauma), other physicians like Haidukewych [11, 12, 13, 14] did not report differences between surgical procedures done within or after twenty-four hours from the injury.

The most recent metanalysis [7, 13, 14] has not proven a direct correlation between time and avascular necrosis occurrence. However, several reports and line guides [15, 16, 17, 18] published in recent years has claimed that it is advisable to treat femoral neck fractures as soon as possible, because of co-factors such as associated lesions, pain control, hospitalization length and cost, and mortality rate at 1 year after trauma.

REDUCTION AND METHODS TO CHECK ITS GOODNESS

For displaced fractures, the first step to manage the lower limb consists of fracture reduction.

An open reduction may be an option if an acceptable alignment cannot be achieved by external manoeuvres, given the importance of appropriate reduction in the rescue of the hip [2]. There are no differences between closed reduction or open capsulotomy (performed in order to reduce hematoma and intracapsular pressure): indeed, no clinical benefits have been demonstrated after capsulotomy: Upadhyay and colleagues [14] did not find any difference in the rates of non-union and avascular necrosis in patients treated either with closed reduction or open capsulotomy

Generally, closed reduction implies the patient positioning on a traction table, with the affected limb adducted and internally rotated; under image-intensified control, the surgeon manipulates the extremity by applying a gentle traction and then he can analyse the result through fluoroscopic images: a first reduction assessment should evaluate the intactness of the posterior cortex, the absence of varus malalignment and the alignment of calcar cortex. In displaced fractures a neck-shaft angle between 130-150°, 10°-15° of valgus and up to 15° of femoral anteversion can be accepted [2], but sometimes imperfect reduction may not be recognize on radiographs. It has been evaluated that together with posterior comminution and improper screws placement, a poor reduction increases the risk of complications [11, 19].

A first method to verify the goodness of a reduction is the Garden alignment index, measured on AP and lateral radiographs; it refers to the angle between the medial cortex and the central axes of the medial trabecular system: in physiologic conditions this is 160° on AP views and 180° on lateral images.

Moreover, the influence of preoperative posterior tilt has been established well: Palm et al. [20] demonstrated that a posterior tilt exceeding 20° in nondisplaced fractures is a reliable predictor of reoperation for implant failure. They measure the angle between the mid-column line and the radius column line (Figure 3).

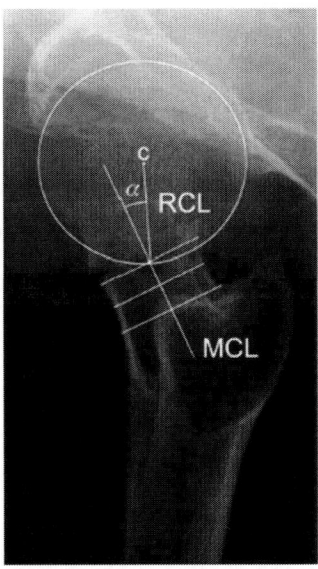

Figure 3. The posterior tilt is measured as the angle (a). Palm et al. [20]

The authors claimed that neither a proper reduction nor the positioning of the fixation implant can compensate for the damage caused by the lateral displacement, probably because of an irreversible damage to femoral vessels and an unstable fracture pattern that is at higher risk of subsequent fracture displacement [20]. Nevertheless Lapidus et al. found no correlation between posterior tilt on lateral pre-operative radiographs and risk for re-operation [21].

TREATMENT

How to Maintain Reduction and Techniques of Fixation

In general, conservative management in young patients is a reliable option where the fragments do not modify their position after the trauma. In practice, there are very few indications for nonoperative management of femoral fracture in young patients due to the significant risk of complications like pseudoarthrosis [1]; moreover, in nondisplaced fractures there is a 20-40% risk of secondary displacement [22, 23]. The risk of avascular necrosis of the femoral head appear to be the same for conservative and operative management (5-11%). For all these reasons, only unfit patient may undergo conservative management.

The main indications for fracture fixation are stable fractures (Garden 1 and 2); even if unstable fractures in young patients should be firstly approach with an attempt of fixation, the total hip arthroplasty is largely used in case of failure [2].

In order to maintain reduction, a three-point contact need to be achieved to contrast the translation of the femoral fragments: the femoral lateral cortex, the subchondral bone of the femoral head, and finally the inferior and posterior aspects of the femoral neck [24].

Extramedullary devices are commonly indicated in non-elderly patients for the treatment of both nondisplaced and displaced femoral neck fractures: the most commonly used implants are represented by cannulated screws and dynamic hip screws, which allow the fragment to slide along the implant with axial loading during weight bearing and consent a controlled dynamic compression during healing [25, 26]. Static external devices are now seldom used because of inferior functional results and higher complications percentages. Biomechanically, the screws resist axial, bending and torsional forces passing through the hip. Regarding cannulated cancellous screws (6,5-7-7,3 mm diameter), several authors studied the effect of orientation and number of the screws to increase stability. The inverted triangle configuration with parallel screws has proven to be more mechanically stable, being characterized by greater torsional and axial stiffness and higher axial loads before failing [19]: the inferior screw resists inferior displacement and its entry point is located just above the lesser trochanter to avoid generating stress onto the subtrochanteric region; the posterior one resist posterior displacement of the fragment. They can be introduced percutaneously and preserve the bone stock better comparing to other implants. The technique to achieve fracture compression is to place the screws perpendicularly to the fracture plane, and parallel one to each other: this may be easier with Pauwels type 1 fracture, where compressive forces predominate. In more vertical fractures shear forces predominate, and placing the screw perpendicularly is more challenging.

The partially threaded screws can be utilized to compress the fracture enhancing healing. A washer to prevent screw penetration in osteoporotic bone is rarely used in young patients but can be a good option in elderly patients.

The fixed angle devices as dynamic hip screw, thanks to their strength to resist varus angulation and inferior fragment displacement, are preferred with Pauwels type 3 fractures, where the fracture line is more vertical, or in basicervical fractures. The fixed angle devices require an open or mini-open surgery. In order to obtain a good implant position, the screw guide wire should be used with an angle guide positioned in the centre of the femoral head in the lateral view and flushed with inferior cortex in the AP view.

Biomechanical studies [27, 28] showed a superiority of DHY over CS in load to failure and fragment displacement; nevertheless, the literature on a clinical comparison between these implants applied in young patients is poor (observational, level I trial studies), and recent metanalysis could not define a clear superiority of DHS over CS. Therefore, further studies are required about this topic.

In Pauwels type 3 fractures, especially if an ipsilateral femoral shaft fracture is present, intramedullary nails may be a valid option to DHS [25, 29]: its medial insertion is an aid for decreasing the deformation forces, and the screw (positioned along the calcar) acts as buttress for the medial cortex of the cephalic fragment.

Arthroplasty

As mentioned before, total hip arthroplasty is nowadays a reliable option in young patients with displaced fractures (Garden type 3 and 4) after a failed fixation: acute arthroplasty carries higher risks than elective procedure, and the good bone stock should be preserved as long as possible, Sometimes, in patients between 65 and 75 years and with initial arthritic signs, THA may be proposed as first option [30, 31]. Several implants are available, with cemented or uncemented stems, modular or fixed neck designs, different combined materials. Despite the higher risk of periprosthetic fracture with uncemented stems, their removal (especially in the context of good bone strength in a young patient) seem to be easier.

Multiple investigators [2] have reported that patients treated with THA after displaced femoral neck fracture have a longer interval to revision surgery or death and improved functional outcomes compared with internal fixation; on the other hand, even if no difference in 30-day mortality rates has been proved there are potentially a higher rate of respiratory complications with THA.

Some authors [2] recognized the potential benefits of THA over other management options for the treatment of femoral neck fractures but emphasized that larger trials are necessary to verify these findings especially in younger patient groups.

COMPLICATIONS

With three cannulated screws, literature reports 90% implant survival rate at 1 year [2]; one disadvantage with their use is the necessity of non-weight bearing after surgery for almost three months, until the healing process has advanced enough to sustain these forces. Among the complications observed, femoral neck shortening (Figure 4) is not uncommon: risk factors are vertical fracture lines (Pauwels type 3 fractures), high BMI, age more than 55; as consequence, worse functional scores have been observed because of increasing lever arm for abductor muscles [32, 33].

In the dynamic hip screws the central screw, with its greater diameter, poses a higher risk of rotational malalignment and AVN for the torque applied to the head fragment during insertion. To reduce these complications sometimes it is advisable to add an anti-rotation screw or wire [2].

As already discussed [25] femoral neck fractures in the young population is commonly associated with other high-energy injuries: the associated biological insult further increases the risk for developing complications, both pre- and post-surgical intervention. The healing potential is reduced by the lack of periosteal layer and limited callus formation because of the growth inhibitors contained in the synovial fluid; these factors lead the risk of delayed healing and non-unions.

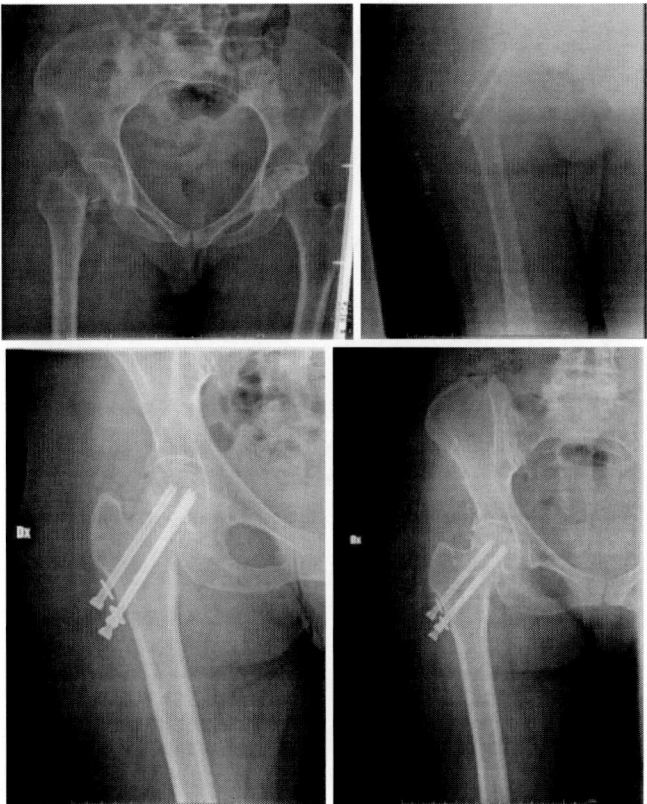

Figure 4. Post-operative images and after 3 months and 1 year in a Garden 4 fracture.

Regarding non-union [19], its prevalence after femoral neck fracture is nowadays 10-20%; it has been correlated with initial displacement, posterior comminution, poor reduction, incorrect screw positioning and non-compressive fixation. Moreover, as mentioned before, the incidence of avascular necrosis is high (7-24% of AVN in fractures treated with internal fixation) [34] and related to several factors other than the direct disruption of the femoral head vascular supply: Haidukewych and colleagues [11] demonstrated an 80% rate of necrosis in mal-reduction cases versus a 24% rate in cases of a good/excellent reduction.

Finally, the risk of requiring an additional procedure increases with inadequacy of reduction, and posterior comminution of the femoral neck: it has been established [34] a 5-15% rate of secondary displacement at 2 years of follow up in young patients treated for displaced femoral neck fractures.

CONCLUSION

Femoral neck fractures in young patients, even if rare, represents a challenging condition for the orthopaedic surgeon. Every attempt should be made in order to salvage the femoral head, and the correct reduction and the subsequent fixation appear to be the keyholes in the management. Cannulated screws in inverted triangle configuration and dynamic hip screws are the most used implants worldwide, but the choice between these devices must be carefully evaluated: the fracture pattern is the major determinant to consider. Regardless the implant chosen and even with a correct reduction, there is the risk of requiring additional procedures due to complications as AVN and non-unions.

REFERENCES

[1]   Goudie, E. B, Duckworth, A. D and White, T. O. 2018 "Femoral Neck Fractures in the Young," in *Proximal Femur Fractures*, Cham: Springer International Publishing: 47–58.

[2]   Florschutz, A. V., Langford, J. R, Haidukewych, G. J. and Koval, K. J. 2015 "Femoral neck fractures: Current management," *J. Orthop. Trauma*, 29 (3): 121–129.

[3]   Garden, R. S. 1971 "Malreduction and avascular necrosis in subcapital fractures of the femur.," *J. Bone Joint Surg. Br.*, 53 (2): 183–97.

[4]   Kang, J. S., Moon, K. H., Shin, J. S., Shin, E. H., Ahn, C. H. and Choi, G. H. 2016 "Clinical results of internal fixation of subcapital femoral neck fractures," *CiOS Clin. Orthop. Surg.* 8 (2): 146–152.

[5]   van Embden, D., Roukema, G, Rhemrev, S. J, Genelin, F, and Meylaerts, S. A. G. 2011 "The Pauwels classification for intracapsular hip fractures: is it reliable?," *Injury*, 42 (11): 1238–40.

[6]   Parker, M. J. and Dynan, Y. 1998 "Is Pauwels classification still valid?," *Injury*, 29 (7): 521–3.

[7]   Gao, Y. S., Ai, Z. S, Zhu, Z. H, Yu, X. W, and Zhang, C. Q. 2013 "Injury-to-surgery interval does not affect postfracture osteonecrosis of the femoral head in young adults: A systematic review," *Eur. J. Orthop. Surg. Traumatol.*, 23 (2): 203–209.

[8]   Swiontkowski, M. F., Winquist, R. A, and Hansen, S. T. 1984 "Fractures of the femoral neck in patients between the ages of twelve and forty-nine years," *J. Bone Jt. Surg. - Ser. A*, 66 (6): 837–846.

[9]   Jain, R., Koo, M., Kreder, H. J, Schemitsch, E. H, Davey, J. R, and Mahomed, N. N. 2002 "Comparison of early and delayed fixation of subcapital hip fractures in patients sixty years of age or less.," *J. Bone Joint Surg. Am.*, 84–A (9): 1605–12.

[10]   Manninger, J., Kazar, G, Fekete, G, Fekete, K, Frenyo, S, Gyarfas, F, Salactz, T and Varga, A. 1989 "Significance of urgent (within 6 h) internal fixation in the management of fractures of the neck of the femur," *Injury*, 20 (2): 101–105.

[11]   Haidukewych, G. J., Rothwell, W. S, Jacofsky, D. J, Torchia, M. E, and Berry, D. J. 2004 "Operative treatment of femoral neck fractures in patients between the ages of fifteen and fifty years," *J. Bone Jt. Surg. - Ser. A*, 86 (8): 1711–1716.

[12]   Popelka, O., Skála-Rosenbaum, J, Bartoška, R, Waldauf, P, Krbec, M and Džupa, V. 2015 " Fracture Type and Injury-to-Surgery Interval as Risk Factors for Avascular Necrosis of the Femoral Head after Internal Fixation of Intracapsular Femoral Neck Fracture" *Acta Chir. Orthop. Traumatol. Cech.*, 82 (4): 282–7.

[13]   Damany, D. S., Parker, M. J, and Chojnowski, A. 2005 "Complications after intracapsular hip fractures in young adults: A

meta-analysis of 18 published studies involving 564 fractures," *Injury*, 36 (1): 131–141.

[14] Upadhyay, A., Jain, P., Mishra, P., Maini, L., Gautum, V. K, and Dhaon, B. K. 2004 "Delayed internal fixation of fractures of the neck of the femur in young adults," *J. Bone Joint Surg. Br.*, 86–B (7): 1035–1040.

[15] Zuckerman, J. D. 1996 "Hip Fracture," *N. Engl. J. Med.*, 334 (23): 1519–1525.

[16] Ftouh, S., Morga, A., Swift, C, and Guideline Development Group. 2011 "Management of hip fracture in adults: summary of NICE guidance.," *BMJ*, 342: d3304.

[17] S. Intercollegiate Guidelines Network. 2009. "Management of hip fracture in older people." *Sign 111*

[18] Smektala R. 2018. "Guidelines or State Civil Codes in the Management of Femoral Neck Fracture? An Analysis of the Reality of Care Provision in North Rhine-Westphalia: In Reply," *Dtsch. Aerzteblatt Online*, 105 (Sbg V): 295–302.

[19] Lowe, J. A., Crist, B. D, Bhandari, M, and Ferguson, T. A. 2010 "Optimal Treatment of Femoral Neck Fractures According to Patient's Physiologic Age: An Evidence-Based Review," *Orthop. Clin. North Am.*, 41 (2): 157–166.

[20] Palm, H., Gosvig, K, Krasheninnikoff, M, Jacobsen, S, and Gebuhr, P. 2009 "A new measurement for posterior tilt predicts reoperation in undisplaced femoral neck fractures: 113 consecutive patients treated by internal fixation and followed for 1 year," *Acta Orthop.*, 80 (3): 303–307.

[21] Lapidus, L. J., Charalampidis, A, Rundgren, J, and Enocson, A. 2013 "Internal Fixation of Garden I and II Femoral Neck Fractures," *J. Orthop. Trauma*, 27 (7): 386–390.

[22] Verheyen, C. C. P. M., Smulders, T. C, and van Walsum, A. D. P. 2005 "High secondary displacement rate in the conservative treatment of impacted femoral neck fractures in 105 patients," *Arch. Orthop. Trauma Surg.*, 125 (3): 166–168.

[23] Shuqiang, M., Kunzheng, W, Zhichao, T, Mingyu, Z, and Wei, W. 2006 "Outcome of non-operative management in Garden I femoral neck fractures.," *Injury*, 37 (10): 974–978.

[24] Della Rocca, G. J. 2015 "Gaps and opportunities in the management of the young femoral neck fracture," *Injury*, 46 (3): 515–518.

[25] Panteli, M., Rodham, P, and Giannoudis, P. V. 2015 "Biomechanical rationale for implant choices in femoral neck fracture fixation in the non-elderly," *Injury*, 46 (3): 445–452.

[26] Pauyo, T. 2014 "Management of femoral neck fractures in the young patient: A critical analysis review," *World J. Orthop.*,5 (3): 204.

[27] Max Hoshino, C. and O'Toole, R.V. 2015 "Fixed angle devices versus multiple cancellous screws: What does the evidence tell us?," *Injury*, 46 (3): 474–477.

[28] Chen, C. Yu, L, Tang, X, Liu, M, Sun, L, Liu, C, Zang, Z, and Li, C. 2017 "Dynamic hip system blade versus cannulated compression screw for the treatment of femoral neck fractures: A retrospective study," *Acta Orthop. Traumatol. Turc.*, 51 (5): 381–387.

[29] Lee, Y. K., Yoon, B. H, Hwang, J. S, Cha, Y. H, Kim, K. C, and Koo, K. H. 2018 "Risk factors of fixation failure in basicervical femoral neck fracture: Which device is optimal for fixation?," *Injury*, 49 (3): 691–696.

[30] Nikiforidis, P., Babis, G. C, Papaioannou, N, Korres, D. S, and Pantazopoulos, T. 1997 "The role of primary total hip replacement for the treatment of the displaced femoral neck fractures," *Eur. J. Orthop. Surg. Traumatol.*, 7 (1): 23–26.

[31] Swart, E., Roulette, P, Leas, D, Bozic, K, and Karunakar, M. 2017 "ORIF or Arthroplasty for Displaced Femoral Neck," *J. Bone Joint Surg. Am.*, 99A (1): 65–75.

[32] Liu, Y., Ai, Z. S, Shao, J, and Yang, T. 2013 "Femoral neck shortening after internal fixation," *Acta Orthop. Traumatol. Turc.*, 47 (6): 400–404.

[33] Zlowodzki, M., Brink, O, Switzer, J, Wingerter, S, Woodall, J, Petrisor, B. A, Kregor, P. J, Bruinsma, D. R, and Bhandari, M. 2008 "The effect of shortening and varus collapse of the femoral neck on

function after fixation of intracapsular fracture of the hip," *J. Bone Joint Surg. Br.*, 90–B (11): 1487–1494.

[34] Scheerlinck, T., and Haentjens P. 2003 "Fratture del femore prossimale nell'adulto," *Surg. Tech. Orthop. Traumatol.*, : 23.

CONTENTS OF EARLIER VOLUMES

INDEX

D

E

R